This textbook will provide both undergraduates and practising engineers with an up-to-date and thorough grounding in the concepts of modern digital transmission. The book is not encyclopaedic; rather, it selects the key concepts and processes and explains them in a deliberately pedagogic style. These concepts and processes are then illustrated by a number of system descriptions.

The book is divided into three parts. The longest, Part 2, describes the basic processes of digital transmission, such as matched filter detection, pulse shaping, line coding, channel coding, error detection and correction, etc. Understanding the concepts behind these processes requires a grasp of basic mathematical models, and this is provided in Part 1. Finally, to put the processes in context, Part 3 describes elements of the public switched telephone network. The text is written throughout in a modern, digital context, and is comprehensively illustrated with helpful figures. Although the mathematical models (time- and frequency-domain concepts) have wider relevance, they are developed specifically for modelling digital signals. The processes described are those found in current transmission systems, and the description of the PSTN includes an outline of newly formulated standards for the synchronous digital hierarchy (SDH), SONET and for broadband ISDN (ATM).

The book will be of great value to 2nd and 3rd year undergraduates studying telecommunications, as well as to graduate trainees and practising engineers. It is appropriate for either private study or as a text associated with a taught telecommunications course. The many worked examples and exercises with solutions will be particularly helpful.

Digital signal transmission

Digital signal transmission

C.C. BISSELL

Faculty of Technology, The Open University

D.A. CHAPMAN

Faculty of Technology, The Open University

Published by the Press Syndicate of the University of Cambridge
The Pitt Building, Trumpington Street, Cambridge CB2 1RP
40 West 20th Street, New York, NY 10011-4211, USA
10 Stamford Road, Oakleigh, Victoria 3166, Australia

First published 1992

Printed in Great Britain at the University Press, Cambridge

A catalogue record of this book is available from the British Library

Library of Congress cataloguing in publication data
Bissell, C. C. (Chris C.)
Digital signal transmission / C.C. Bissell, D.A. Chapman.
p.   cm.
Includes bibliographical references and index.
ISBN 0–521–41537–3. – ISBN 0–521–42557–3 (pbk.)
1. Digital communications.   2. Signal processing – Digital
techniques.   3. Integrated services digital networks.   I. Chapman,
D.A. (David A.)   II. Title.
TK5103.7.B54   1992
621.382 – dc20   91/40848   CIP

ISBN 0 521 41537 3 hardback
ISBN 0 521 42557 3 paperback

Contents

Preface

In this book we try to give a representative (but not comprehensive) treatment of the digital transmission of signals. Our main aim has been to render the material truly accessible to second or third year undergraduate students, practising engineers requiring updating, or graduate physicists or mathematicians beginning work in the digital transmission sector of the telecommunications industry. This has led to a book whose important features are:

> A limited number of topics, dealt with in depth
>
> An emphasis on the engineering context and interpretation of mathematical models
>
> Relevance to both students and practising engineers

Engineering is a pragmatic activity, and its models and theory primarily a means to an end. As with other engineering disciplines, much of telecommunications is driven by practicalities: the design of line codes (Chapter 6), or the synchronous digital hierarchy (Chapter 8), for example, owe little to any complicated theoretical analysis of digital telecommunications! Yet even such pragmatic activities take place against a background of constraints which telecommunications engineers sooner or later translate into highly abstract models involving bandwidth, spectra, noise density, probability distributions, error rates, and so on. To present these vitally important ideas – in a limited number of contexts, but in sufficient detail to be properly understood by the reader – is the main aim of this book. Thus time- and frequency-domain modelling tools form one constant theme (whether as part of the theory of pulse shaping and signal detection in Chapter 4, or as a background to the niceties of optical receiver design in Chapter 9); the constant battle against noise and the drive to minimise errors is another.

Chapter 1, and the brief introductions to the three main parts, describe the scope of the book more fully. We believe there are already enough encyclopaedic reference texts and books of applied mathematics masquerading as introductions to telecommunications! We have tried to write a book in which the theory illuminates, rather than obscures, the engineering; and a book which will enable the reader subsequently to approach more advanced reference material (specialist texts and manufacturers' literature, for example) without being overwhelmed.

Acknowledgments

Large parts of this text are derived from our contributions to the Open University course T322 Digital Telecommunications (Open University Press, 1990), although the material has been considerably modified, updated, and extended. Our general approach has been greatly influenced by discussions within the T322 course team, and in particular by the stimulating ideas of the Course Chair, Gaby Smol. Around 60% of the figures used here are reproduced directly from T322 by kind permission of the Open University. Thanks are also due to Mr P.J. King of GPT and Dr V.S. Shukla of GTE for advice on the contents of Chapter 8. Any errors, however, remain our own responsibility.

Some of the text was written while one of us (Bissell) was on study leave at Twente University in the Netherlands; the hospitality and accommodation provided by the Vakgroep Tele-Informatica is gratefully acknowledged.

Chris Bissell
David Chapman
The Open University

1

Introduction

The topic of digital signal transmission is an enormous one, and cannot be covered completely in a book this size. The aim of this introduction is to set the material of later chapters into the general context of modern telecommunication systems, and to outline what will be covered in detail later and what will not. Some suggestions will also be made about how to use the book.

1.1   The Integrated Services Digital Network

In most countries, telecommunications services are evolving rapidly from a collection of separate, and largely incompatible, systems (telephone, telex, public and private data networks, and so on) towards a universal network, in which a wide variety of services are integrated using a common (digital) form of transmission.

Fig. 1.1 shows, in much simplified form, how two offices (or factories, or homes) might be interconnected through such an *Integrated Services Digital Network* (*ISDN*). At each office, digital signals from a number of different audio, video or data terminals are connected, via appropriate network terminating equipment (NTE), to an exchange termination (ET) within the ISDN. Whatever their origin, the signals are transmitted to their destinations over digital links which may include optical fibres, metallic cables, terrestrial or satellite microwave channels, and so on. To the network it is completely immaterial whether the signal carries a telephone conversation, computer data, or the reading on an electricity meter.

A universal network should eventually prove more flexible and cheaper than a number of separate networks, each providing only a narrow range of services. Ultimately such a network may even subsume broadcast radio and television! However, existing telecommunications networks represent enormous investment, and cannot be simply abandoned and replaced by an ISDN. Evolution towards a universal network is therefore gradual, and depends upon the adaptation of existing systems.

Chapter 8 will look in detail at some aspects of the evolution of existing telephone networks towards an ISDN.

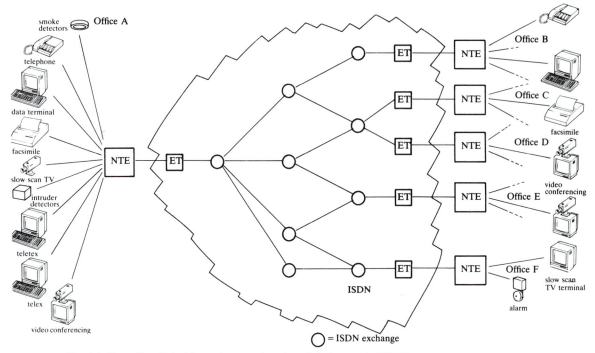

Fig. 1.1. Two offices linked by an integrated services digital network (ISDN).

1.2   Standards

A moment's reflection on the nature of an ISDN should bring home the importance of standards for the success of such a multi-purpose, wide-ranging system. Standards are required at all levels – from internationally agreed regulations on the interconnection of national ISDNs, to the details of voltages and wiring for the connection of individual terminals to NTEs.

Standards have long been vital for the telecommunications industry, and many sets of national and international regulations have been published. Two of the most important bodies are the *CCITT* (French initials for International Telegraph and Telephone Consultative Committee – part of the International Telecommunications Union (ITU), affiliated to the United Nations Organization) – and the *ISO* (International Organization for Standardization). CCITT recommendations are issued approximately every four years, as a set of volumes universally referred to within the telecommunications industry by their colour: the current set, published in 1989, are the *Blue Books*. Many references to CCITT recommendations will be made in subsequent chapters.

1.3 Convergence of telecommunications and computing

Implicit in Fig. 1.1 is a comparatively recent phenomenon known as the 'convergence' of digital telecommunications and computing. As the use of computers in business, industry and even the home to store and process information has grown, so there has been an increasing demand for the electronic transmission of such information – both within a single organisation via local networks, and from one organisation to another over public and private networks.

At the same time, most recent advances in telecommunications have involved digital techniques for both the transmission and processing of signals. As far as transmission is concerned, the reasons include:

- the fact that the build-up of noise on a conveyed message can be virtually eliminated by the process known as regeneration (outlined below), and the error rate can thus be kept very low
- the possibility of detecting and even correcting those errors which do occur during signal processing and transmission (discussed in Chapters 5 and 6)
- the development of high capacity transmission media, such as optical fibres, which are particularly suited for digital transmission

Part II of this book is concerned mainly with such properties of digital signals, and with techniques for processing, transmission and error detection/correction.

Furthermore, recently developed digital electronic devices are cheaper, more reliable and consume less power than the analogue circuits previously used for signal processing.

1.4 Open Systems Interconnection

The convergence between telecommunications and computing has led to a need for standards which deal with much more than simply the conveying of a digital signal from transmitter to receiver at a suitably low error rate. Systems which offer, say, facilities for processing data on one computer using programs accessed remotely on another, or which allow complex graphics to be transmitted from one make of personal computer to another in a different country, clearly require a degree of standardisation quite unlike that traditionally associated with the telecommunications industry! An important strategy has emerged over the last decade or so to deal with such complexity, based round the ISO Open Systems Interconnection (OSI) Basic Reference Model. Whole books have been devoted to the OSI reference model, so the description which follows is

See, for example, MacKinnon, McCrum & Sheppard.

necessarily an over-simplification. It is included for two reasons: first, because it has become standard practice for descriptions of modern digital telecommunication systems to refer to the OSI model, so the subject matter of this book needs to be placed in this general context; second, because one of the central features of the model – the concept of *layers* – is used as a framework for Part 2 of this book.

The origins of the OSI reference model in computer communications is evident in some of the terminology used. Thus the term *application* is used to refer to the complete set of processes involved in a particular customer service (rather as in 'applications program'): electronic mail or airline ticket reservation are examples of applications in this sense. A system is called *open* if it is capable of communicating directly with other open systems of an appropriate type. An open system conforming to OSI standards carries out all communication processes for the applications it serves in a standard, agreed way, so it can work with other open systems running cognate applications. For example, a particular airline might run Application A to handle its own flight bookings, while a travel agency runs Application B dealing with bookings with a number of different airlines. OSI is not concerned with the booking process itself (this is a function of the applications), but rather with the communication systems. The latter consist mainly of software processes which allow the airline and travel agency applications to interact.

Perhaps the most distinctive feature of the OSI reference model is its layered architecture, in which a hierarchy of abstraction is identified in the overall communication process. This is shown in Fig. 1.2. The precise functions of each individual layer are not a major concern of this book, but can be briefly summarised as follows:

> *physical layer* – does whatever is needed to convey bit streams between the layer immediately above it and the physical trans-mission medium. This may include the use of a modem or other signal processing to produce a physical signal appropriate to the medium.
>
> *data link layer* – ensures the correct, effectively error-free trans-mission of data between individual nodes in the network.
>
> *network layer* – deals with the routing and switching of the data over the complete network, including the orderly re-establishment of connection in the case of disruption.
>
> *transport layer* – sets up end-to-end connections, possibly selecting from more than one network if available.
>
> *session layer* – organises the whole set of transport connections

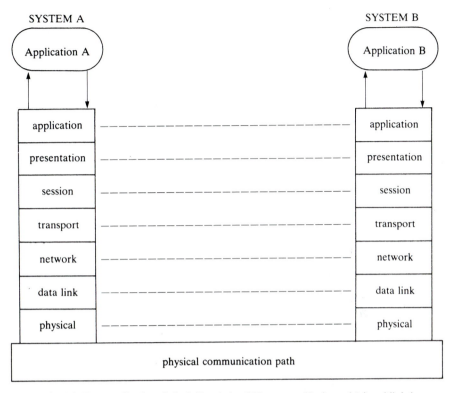

Fig. 1.2. Two applications linked directly by OSI systems. Horizontal 'virtual links' can be set up between any pair of 'peer' layers.

required for a given session; carries out various other functions relating to data interchange.

presentation layer – deals with the rules for representing information, agrees a common syntax such as the use of ASCII code.

application layer – deals with the semantics of the communication, normally specific to the particular application. In the ticket reservation example, a common language for referring to flight numbers, times, and so on, would be required.

In the OSI model, communication takes place vertically as requests for action are passed from one layer to another. Suppose, for example, that the system A transport layer of Fig. 1.2 requests from the network layer a connection to system B. The network layer passes on an appropriate request to the data link layer, and the data link layer to the physical layer. A suitable message is then transmitted over the physical communication path, and ultimately the request for a connection reaches the network layer of system B. To the two network layers, however, it is as though there is a direct horizontal communication channel – known as a *virtual channel* – between them. This is an important feature of the OSI model,

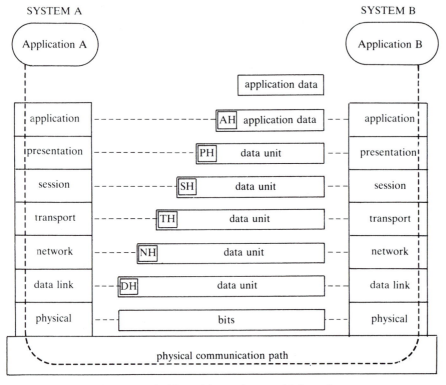

Fig. 1.3. Vertical flow of data and protocol information.

and is represented by the dashed lines in Fig. 1.2. Horizontal communication over virtual channels is specified in terms of so-called *peer protocols*. The protocol information required by the individual layers is transmitted as a set of *headers* added to the application data on its way down one stack and stripped off on its way up the other, as shown in Fig. 1.3. The price to be paid for the flexibility of OSI standardisation is thus the transmission of a large data overhead.

In OSI terminology, this book is almost entirely concerned with the physical layer – that is, with techniques used to transmit digital signals reliably over particular communication paths. Some aspects of Parts 2 and 3, however, are related to higher-layer OSI functions. The general notion of a layer model, in which each layer adds an overhead as the price of some desirable feature, will also reappear in Part 2.

1.5   Analogue and digital signals

The term 'digital signal' has been used so far without definition, on the assumption that the reader will have some familiarity with digital systems.

Now, however, it is time to be more precise, and to turn to digital signal transmission in detail. For the purposes of this book, a digital signal is one which consists of a sequence of symbols taken from a finite set. The text of this book can therefore be thought of as a digital signal: it is made up of a finite number of distinct symbols – the Roman alphabet (and part of the Greek), in various founts and formats, together with punctuation marks (including spaces), the digits 0 to 9, and mathematical symbols.

The simplest digital signal – a *binary signal* – uses only two symbols or states, almost always denoted 0 and 1. For transmission over a physical communication path, these binary digits or *bits* must be represented in some appropriate way. A number of typical binary signals are shown in Fig. 1.4. In parts (a) and (b) the binary states are represented by two different voltage levels in a way known as *positive logic* if the voltage used to represent binary 1 is higher than that used for binary 0, and *negative logic* if the converse is true. The pulses shown are *non-return to zero* (*NRZ*): each voltage level is held at the appropriate value for the whole

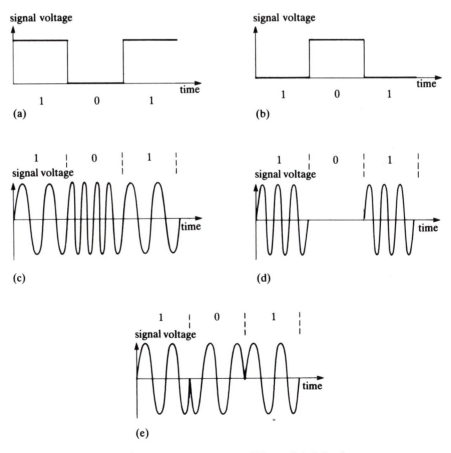

Fig. 1.4. Some common types of binary digital signals.

symbol period. (A common alternative is for the non-zero voltages to last for only part (usually half) of the symbol period: this is *return-to-zero* (*RZ*) signalling.) The waveforms of parts (a) and (b) are both *unipolar* – voltages of one polarity only are used, in contrast to *bipolar* (or simply *polar*) signalling, where both positive and negative voltages are used in a single digital waveform. In part (c) of the figure, two different frequencies are used to represent the binary states, while in (d) a single frequency is turned on and off. Finally in (e) the phase of a transmitted sinusoid carries the information.

The terminology used to describe digital waveforms is not standard, and readers should be on their guard against confusion. For example, in North America the term bipolar has a different meaning, as will be described in Chapter 5.

   One of the most important features of digital signals is the way in which the effects of noise and interference can be virtually eliminated. Fig. 1.5 illustrates the general principle. Part (a) shows a binary signal representing the bit stream 10110001. After transmission over a noisy, distorting channel, the received signal might be as shown in part (b). Although the signal has a very different appearance from that transmitted, it is possible to obtain a perfect replica of the original by means of a circuit which

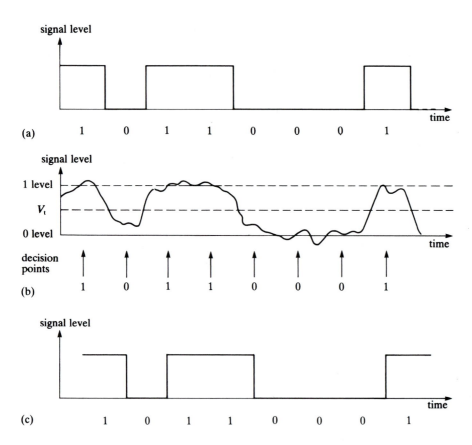

Fig. 1.5. Binary waveforms: (a) transmitted, (b) received and distorted, (c) after regeneration.

samples the distorted waveform at the mid point of the symbol period, and generates a high or low voltage output depending on whether the sample is above or below an appropriate threshold voltage V_t. The process is known as *regeneration*, and the result is shown in part (c) of Fig. 1.5. Of course, if the noise and distortion are sufficient for the received signal to take on a value the wrong side of the threshold at the sampling instant, a digital symbol can be misinterpreted and an error will occur. Nevertheless, by locating *regenerative repeaters* at suitably spaced intervals along a link the error rate in a digital transmission system can be kept very small indeed.

Many signals which need to be transmitted will already be in digital form – computer data or text files, for example. Many other signals, however, are not naturally restricted to a finite set of states, but can vary continuously as an analogue of some physical quantity – a telephone speech signal, for example, or signals representing temperature, pressure, or other variable in a telemetry application. It is now very common for such *analogue signals* to be processed and transmitted digitally. In order to do this, the analogue signal is sampled, and the sample value rounded to the nearest of a finite set of quantisation levels. Each level can then be allocated a binary code, and the original signal becomes a sequence of code words which can be transmitted using normal digital techniques.

> **Sampling is discussed in Chapter 2, while the particular system used in telephony (*pulse code modulation*) is considered in detail in Chapter 7.**

The elaboration of the previous few paragraphs forms the major subject matter of this book. Typical questions which will be addressed in detail are:

- what are the best digital symbols to use for a given communications channel?
- how can the effects of noise and distortion be modelled, quantified and combatted?
- how can an accurate timing signal be ensured (as implied in Fig. 1.5), so that regenerators and receivers sample close to the ideal instant?
- if errors do occur, how can they be detected and corrected?
- what sort of additional information needs to be included with the data to facilitate timing extraction and error correction?

1.5.1 Switching

Before turning to these topics, it is worth briefly mentioning one other important aspect of digital telecommunications. Consider again Fig. 1.1, the nodes of which represent ISDN exchanges capable of switching any incoming channel to any outgoing one in order to connect any two users. On a national scale, switched networks need to be able to interconnect any two users (or more, for services such as telephone- or video-conferencing) out of millions of subscribers. Switching centres (exchanges) for such networks are based on computer-controlled, digital techniques, and their design is a major undertaking involving many years of engineering time. Although switching will not be considered in this text, it is worth noting a distinction commonly made between *circuit switching* (as in the public switched telephone network) and *packet switching* (used widely for electronic mail and data communication in general). In circuit switching, end-to-end links are set up for the duration of a call. The links remain reserved for the call even if unused during temporary lulls in transmission. The overall transmission delay is the propagation time through the network – less than a few tenths of a second even if satellite links are involved. In packet switching, messages are broken up or combined as appropriate into packets of convenient length. Packets are sent individually, stored at nodes if necessary until a link is free, and reassembled at the destination. The time taken depends on the number of nodes involved, the levels of traffic, and the packet switching technique used: it may be a few seconds (or even less), or up to several hours for a crowded international route. Since packets are only sent when needed, and the links are otherwise available for other connections, packet switching improves the efficiency of the network – at the expense of increased transmission delays due to processing and, possibly, storage at intermediate nodes. As packet switching becomes faster and more efficient, however, the distinction between it and circuit switching is beginning to blur, as will be noted in Chapter 8 in the context of a recent development known as *asynchronous transfer mode (ATM)*. The details of the control and supervision of both circuit- and packet-switched networks lie outside the scope of this book, being the province of higher levels of the OSI reference model.

1.6   How to use this book

This book is divided into three parts. *Part 1: Models* introduces those random and deterministic models of signals and systems which have proved to be of immense value in the design and analysis of telecommunication systems. In contrast to most other books on digital telecommunications, the mathematical formalism has been kept to a minimum, with references to other standard works being given where appropriate for detailed proofs and analytical background. It is the authors' belief that engineers need only limited expertise in carrying out mathematical operations, but that they need to understand fully the significance and implications of the models they use. With this in mind, more attention has been paid to explaining modelling assumptions in full and discussing the engineering implications of the models used than to listing all the fine details of the mathematics itself. Part 1 should thus serve as either an introduction to signals and systems, or as revision – perhaps from a rather different viewpoint – for those who have come across the material before.

Those with some prior knowledge may prefer to begin immediately with *Part 2: Processes*, which applies the material of Part 1 to the specific task of digital signal transmission. Numerous cross-references to earlier material are provided as an aid to such readers. Part 2 is based upon a layer model of digital transmission, with successive chapters devoted to pulse transmission, line coding and channel coding. A restricted number of topics are covered in depth, instead of the more comprehensive, but highly condensed, treatment common in other texts.

Part 3 looks in detail at the public switched telephone network (PSTN), taking up again some of the topics briefly discussed in this introductory chapter in the light of the detailed development of Parts 1 and 2. While in no sense an exhaustive treatment of the PSTN, Part 3 sets many of the concepts of Parts 1 and 2 firmly in an engineering context. A hierarchical approach is adopted: first a look at techniques for encoding analogue signals and the use of modems for digital sources; then some details of approaches to transmission over the digital network; and, finally, the design of a single optical fibre link within the PSTN. The wide range of topics and concepts in this part is deliberate, and designed to reflect the breadth demanded of a modern telecommunications engineer. Part 3 is also reasonably self-contained, with cross-references to earlier parts of the book.

The prior knowledge required by the reader is modest, and should be covered by most first-year engineering curricula – and even many A-level syllabuses. An understanding of complex numbers and elementary calculus is essential, as are basic notions of frequency response and the behaviour of simple RC networks. Some exposure to circuit theory, digital devices (logic gates and shift registers), and operational amplifiers would be useful, but is not absolutely necessary. Worked examples and exercises (with outline solutions) are provided throughout to assist those readers using the book as a self-study text.

Part 1
Models

The two chapters of this part are devoted to models of signals, systems and noise. Such models have long been vital tools of the telecommunications engineer: many of them, in fact, date back to the early days of telegraphy, although they are applied in very different ways to modern telecommunications systems. A number of themes running through the next two chapters can be identified. First, there is the important notion of equivalent descriptions of signals as waveforms in the *time domain* or as spectra in the *frequency domain*. Secondly, there is the contrast between signals which are specified in advance for all time – the *deterministic* signals of Chapter 2 – and those *non-deterministic* or *random* waveforms, such as noise, which require the statistical techniques introduced in Chapter 3. Finally, there is an emphasis on the way engineers – rather than mathematicians or system theorists – use these models in the design and analysis of digital telecommunications systems.

As mentioned in Chapter 1, mathematical theory has been kept to a minimum, and a number of standard mathematical results have been collected as Appendices A and B rather than included in the main text. The reader is also referred to other standard works where appropriate. This is not to detract from the importance of mathematics for digital telecommunications, but rather to put the emphasis where it belongs: on a deep understanding of the functions and limitations of the models used, and on the need to interpret such models in their engineering context.

2

Signals and systems

2.1   Introduction

Fig. 2.1 introduces some of the most important phenomena which need to be modelled and analysed in digital signal transmission. It shows part of a system for transmitting binary data coded as two different voltage levels: a simple non-return-to-zero (NRZ) code has been assumed, in which a binary 1 is represented by a positive voltage for the duration of a complete symbol period, and binary 0 is represented by zero volts. The transmission medium might be a coaxial cable, as often used in small scale local networks. Similar principles apply, however, to systems using other transmission media and/or more complicated codes.

Fig. 2.1 also shows (not to scale) typical waveforms at various points in the system. After passing down the cable the original waveform A is attenuated (by an amount depending on the length of the cable) and, because of a finite system response time, the clear transitions between the two voltage levels have become indistinct. In practice there will also be a delay, corresponding to the time taken for the signal to pass along the cable, although this has not been shown explicitly. Neither are the effects of any noise included.

To counteract the distortion illustrated, the system includes an *equaliser*, which 'sharpens' the received waveform, so that the relationship

This figure can be thought of as an elaboration of some aspects of Fig. 1.5 of Chapter 1.

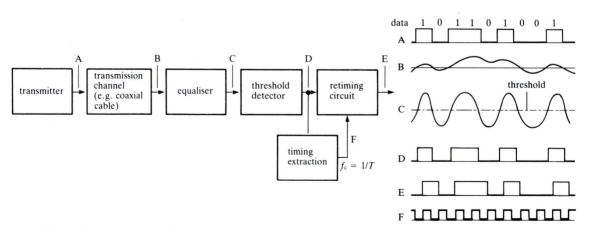

Fig. 2.1. Part of a binary digital transmission system, showing typical waveforms at various points.

of the equaliser output C to the original binary symbols is much clearer. Passing this equalised waveform through a *threshold detector* (a circuit whose output is one of two voltage levels depending on whether the input is greater or less than a pre-set threshold) has the result of generating a binary signal very similar to the transmitted one. In fact, providing the noise levels are sufficiently low, and the equaliser and threshold detector are properly designed, then the only difference between binary waveforms A and D is that the transitions of the latter are not perfectly in step with those of the former. The transitions of D will correspond to the threshold-crossings of waveform C which, even in the best practical system, will not precisely mirror those of the original binary waveform.

The final stage of Fig. 2.1 is the retiming of the received waveform. If this were not carried out, then the irregularities (*jitter*) in waveform D would soon build up to cause error over a long link. A regular timing reference signal F – often known as the clock – is derived from the received waveform itself by a special circuit. The clock signal and the output from the threshold detector are then processed to give a final *regenerated* digital signal E whose transitions now coincide with the instants at which the clock signal goes from low to high. A comparison of waveforms C and E shows that the combined effect of threshold detection and re-timing is equivalent to sampling waveform C near its peaks and troughs to determine the appropriate binary states. So even in the presence of noise, the regenerated signal E can be an almost perfect (delayed) replica of the transmitted signal, providing only that the noise is not sufficient to cause an incorrect decision to be made at the threshold detector.

From this brief discussion it should be clear that to design and analyse such systems we need a set of appropriate modelling tools, above all for evaluating the effects of the various system components on the precise shape of the received digital symbols.

2.2   Linearity

A particularly important type of mathematical model used in science and technology is the *linear model*. As used in telecommunications engineering, the term linear implies rather more than a simple straight-line relationship between a system's input and output. Of the several equivalent definitions of linearity, the following two will concern us here:

1 A linear system is one which obeys the principle of *superposition*. That is, if an input $x_1(t)$ produces an output $y_1(t)$ and an input $x_2(t)$ produces an output $y_2(t)$ then an input $x_1(t) + x_2(t)$ produces an output $y_1(t) + y_2(t)$.

2 A linear system is one possessing the 'frequency preservation' property. That is, the steady state response of a linear system to a sinusoidal input is itself a sinusoid of the same frequency, but generally differing in amplitude and phase.

An extensive set of mathematical tools and practical techniques has been developed to handle linear systems, and many elements in a telecommunication system are designed to be as nearly linear as possible. A sufficiently close approximation to linearity means, for example, that the principle of superposition can be used to model the effect of a channel or other system element on a sequence of pulses, given a knowledge of the effect on one pulse. Fig. 2.2 illustrates this for a system element which

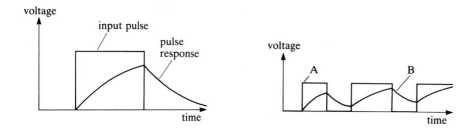

Fig. 2.2. (a) Rectangular pulse response of a first-order lowpass filter, where the duration of the pulse is approximately equal to the filter time constant. (b) Response of the same filter to a binary waveform.

behaves like a first-order lowpass filter. Part (a) shows the response to a single pulse, while part (b) shows the superposed pulse responses corresponding to an input pulse train (binary waveform). Note that because the response to a single pulse takes longer to decay than the duration of a symbol period, the output waveform B gradually accumulates an offset. In the absence of further processing this would clearly cause problems for threshold detection and regeneration of the original binary signal: even in the positions corresponding to a binary 0 there is a considerable output voltage. Fig. 2.3, on the other hand, shows a much more desirable overall pulse response for a telecommunications channel. Here the system response to a bit stream could be decoded without difficulty, owing to the clear distinction in the combined response between binary 1 and 0. This topic will be discussed in much more detail in later chapters.

An alternative approach to modelling a linear channel or component is based on the second definition of linearity. As will be discussed in detail below, any practical message signal can be described in terms of its

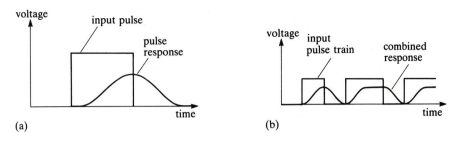

Fig. 2.3. Possible response of a telecommunications channel to (a) a rectangular pulse and (b) a binary waveform.

Not all elements in a telecommunications system are designed to be linear. For example, it is often necessary to *change* the frequency of a signal and any device doing this is, by definition, non-linear.

frequency content – or, to be more precise, modelled as a frequency spectrum. Similarly, any linear system can be completely specified by its frequency response function: a description of amplitude and phase shifts introduced by the system for all frequences.

Frequency response functions are often written using the notation of complex numbers. For example, the frequency response function $H(f)$ of a first-order lowpass system can be written

$$H(f) = \frac{k}{(1 + \mathrm{j}f/f_c)}$$

where k is the low-frequency or d.c. gain; f_c is the 3 dB cut-off frequency; and $\mathrm{j}^2 = -1$. In general, $H(f)$ takes on a complex value for any given values of k, f and f_c.

It is assumed that readers have met the notion of frequency response and the use of complex numbers to represent the amplitude and phase of sinusoids. An introduction can be found in Meade & Dillon.

For a linear system with a frequency response function $H(f)$ an input sinusoid $x(t) = A \sin \omega t$ produces a steady-state output $B \sin(\omega t + \theta)$ where the ratio $B/A = |H(f)|$ is known as the *amplitude ratio* and $\theta = \operatorname{Arg} H(f)$ is the *phase shift*. Frequency response functions are usually plotted as separate graphs of such amplitude ratio and phase shift, as shown in Fig. 2.4 for a first-order lowpass system: a simple *RC* filter, perhaps, or an operational amplifier circuit. Note that, as is often the case, the amplitude ratio is expressed in decibels – that is, $20 \log |H(f)|$ and that the figure has been drawn assuming a d.c. gain k of 1 (0 dB). Simple scaling allows the curve to be used for other values of k.

2.2.1 Amplitude distortion and phase distortion

An ideal transmission channel would pass all frequency components of a signal with their amplitude and phase relationships unchanged. The simplest frequency domain model of such behaviour would be a constant amplitude ratio and zero phase shift for all frequencies of interest. Then

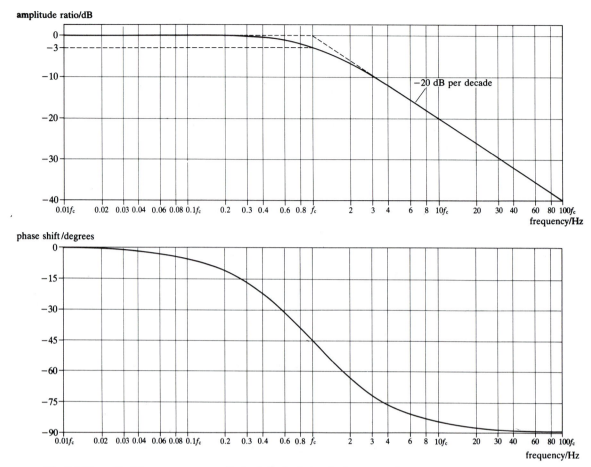

Fig. 2.4. Normalised frequency response (Bode plot) of a first-order lowpass system.

the received signal would be a perfect (scaled) replica of the transmitted signal, as illustrated in Fig. 2.5(a), where the scaling factor B/A represents the constant amplitude ratio of the channel. But zero phase shift for all frequencies implies zero propagation time, which is hopelessly unrealistic. The next best thing (and even this turns out to be unrealisable in practice) would be for all sinusoidal components – and hence the signal as a whole – to be delayed by exactly the same length of time, as shown in Fig. 2.5(b). The signal profile would then be unchanged (provided the channel amplitude ratio is constant as before), but there would be a constant propagation delay t_D for all frequencies.

If all sinusoids are delayed by t_D then a transmitted $A \sin \omega t$ of arbitrary frequency will be received as $B \sin[\omega(t - t_D)] = B \sin(\omega t - \omega t_D)$. The term $-\omega t_D$ represents the phase shift θ, so the channel introduces a phase *lag* proportional to frequency; such a phase characteristic is as shown in Fig. 2.6 and is known as a *linear phase* characteristic.

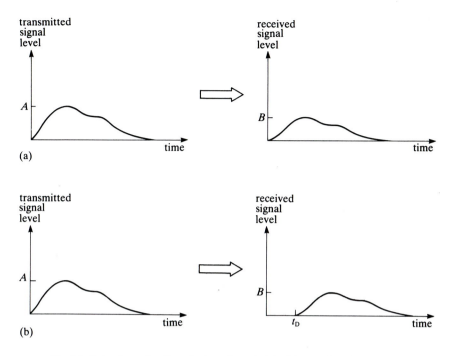

Fig. 2.5. Behaviour of a distortionless channel: (a) with zero time delay; (b) with a constant transmission delay for all frequencies of interest.

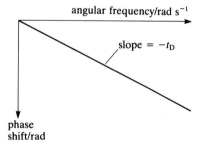

Fig. 2.6. A linear phase characteristic corresponding to a transmission delay of t_D seconds.

A channel with an amplitude ratio which is not constant for all frequencies of interest is said to introduce *amplitude distortion*; a channel whose phase characteristic is not linear introduces *phase distortion*.

One important feature of linear phase is illustrated in Fig. 2.7, which shows possible rectangular pulse responses of channels which (a) introduce both amplitude and phase distortion and (b) amplitude distortion alone. In both cases the original pulse is distorted, but note that the received pulse is *asymmetric* when both amplitude and phase

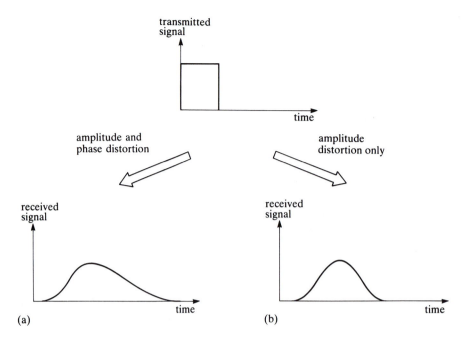

Fig. 2.7. Possible distortion of a rectangular pulse: (a) both amplitude and phase distortion; (b) amplitude distortion alone.

distortion are introduced, but *symmetric* for linear phase channels, when only amplitude distortion takes place. For digital systems in which symmetric pulses are transmitted in regular time slots, a linear phase characteristic is a highly desirable property, since the symmetry of the received pulses makes it easier to prevent a pulse in one time slot spreading out into adjacent ones.

Contrast, for example, the amplitude and phase distortion inherent in Fig. 2.2 with the symmetric pulse response (linear phase) implied in Fig. 2.3.

2.3   Frequency-domain representation of periodic signals

A frequency response model of a telecommunications channel is a very useful tool, which can give an experienced engineer an immediate 'feel' for how the channel will behave in general terms, without the need for complex calculations. To be able to use such frequency domain models, however, it is necessary to characterise a digital message signal, such as the pulse trains considered above, in terms of their frequency content or spectra.

2.3.1   Fourier series

In 1822 Jean Baptiste Joseph Fourier derived a classic result which has proved to be of enormous significance for many branches of science and technology. One way of stating his theorem is as follows:

Any periodic signal $f(t)$ can be expressed as a series, possibly infinite, of sinusoidal components, such that

$$f(t) = A_0 + A_1 \cos(\omega t + \phi_1) + A_2 \cos(2\omega t + \phi_2) + A_3 \cos(3\omega t + \phi_3) + \dots$$

where the fundamental angular frequency ω of the *Fourier Series* is related to the repetition period T_p of the signal $f(t)$ by the expression $\omega = 2\pi/T_p$, and the Fourier coefficients A_n and ϕ_n are constants.

The coefficient A_0 represents the d.c. component or mean value of the signal, while $A_1, A_2 \dots A_n$ represent the amplitudes of the sinusoidal components. The phase coefficients $\phi_1, \phi_2 \dots \phi_n$ represent the phase angles of these sinusoids at the various multiples of the fundamental frequency. A Fourier series can be written in a number of equivalent ways: as a cosine series, as a sine series, as a combination of sine and cosine terms, or in exponential form. Some standard results of Fourier series are listed for reference in Appendix A.

From a mathematical point of view, a periodic signal is eternal, extending from $-\infty < t < +\infty$. What this means in practice is that the signal has existed in its periodic form sufficiently long for any start-up transients introduced by switching on to have completely died away.

For the engineer, the important consequence of the above result is that any periodic waveform can be treated exactly as though it consisted of the set of sinusoids in its Fourier series representation. The effect of a linear system element on each of these sinusoids can be easily calculated from a knowledge of the system's frequency response.

Suppose, for example, that the binary pattern $\dots 10101010 \dots$ is to be transmitted using $\pm V$ to represent the two binary states. The transmitted signal would therefore approximate to the square wave of Fig. 2.8, where the waveform has been drawn symmetrically about the time origin for mathematical convenience.

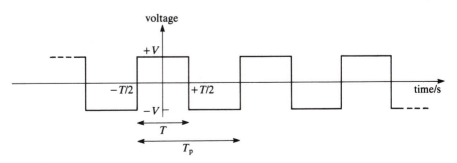

Fig. 2.8. A square wave representing $\dots 101010 \dots$.

The Fourier series corresponding to Fig. 2.8 is

$$f(t) = \frac{4V}{\pi} \left[\cos \omega t + \frac{1}{3} \cos (3\omega t + \pi) + \frac{1}{5} \cos 5\omega t + \frac{1}{7} \cos (7\omega t + \pi) + \dots \right]$$

where again $\omega = \dfrac{2\pi}{T_\mathrm{p}}$.

Note that for this particular waveform only odd harmonics – that is, only odd multiples of the fundamental frequency $1/T_\mathrm{p}$ – are present, and that the phase angles of the components are alternately 0 and π rad (0 or 180°). Note also that there is no d.c. term A_0: the d.c. term represents the mean value of the waveform which, in this case, is clearly zero. The repetition period of the square wave is twice the duration of a signalling element, so $T_\mathrm{p} = 2T$ and the fundamental frequency (in hertz) is $\frac{1}{2T}$.

A Fourier series can be represented diagrammatically as a line spectrum, showing the amplitude and phase of all components. Fig. 2.9 shows separate amplitude and phase spectra corresponding to the signal of Fig. 2.8. (Note that the frequency axis is labelled in hertz rather than radians per second for convenience.) In general two, separate, amplitude

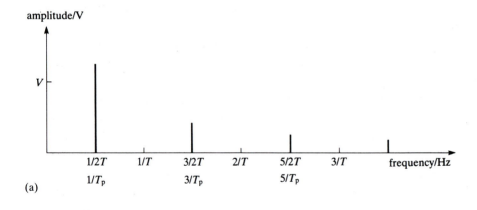

(a)

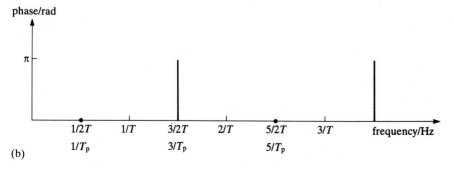

(b)

Fig. 2.9. The amplitude and phase spectra of the square wave of Fig. 2.8.

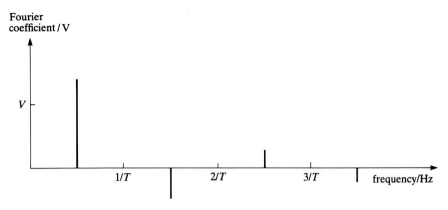

Fig. 2.10. An alternative representation of the line spectrum of Fig. 2.9.

and phase plots will be necessary to specify a spectrum, but for Fourier series like this one, in which the phase angles are all 0 or $\pm\pi$ rad, a simplification is possible. Since a phase shift of $\pm\pi$ is equivalent to multiplying a sinusoid by -1, $\cos(\omega t \pm \pi)$ can be written as $-\cos\omega t$. Hence the anti-phase components of the series under consideration can be represented as negative lines on the spectral diagram, as shown in Fig. 2.10. A waveform which is symmetric about the time origin, such that $f(t)=f(-t)$, is known as an *even signal*. It can be shown that the phase angles of the cosine series corresponding to any even signal are restricted to 0 or $\pm\pi$ (or multiples thereof, which amounts to the same thing), and so the spectra of such even waveforms can always be represented on a single diagram.

An important point to note about the spectrum of the square wave of Fig. 2.8 is that significant frequency components extend to very much higher frequencies than the fundamental frequency. To transmit the waveform without significant distortion would therefore require a channel with a considerable bandwidth, not to mention a suitable phase characteristic. Of course, as was illustrated earlier, a pulse train does not have to be received undistorted in order to make correct decisions about the binary states. In fact, in the case illustrated, so long as the fundamental component at $1/(2T)$ Hz of the square wave corresponding to the bit stream $\ldots 10101010\ldots$ can be transmitted, then correct decisions can be made about the binary states. This is illustrated in Fig. 2.11, showing how it is possible, in theory, to transmit $1/T$ symbols per second over a channel of bandwidth $1/(2T)$ Hz. Put another way, it is possible to signal at a rate of $2B$ symbols per second over an ideal bandlimited channel of bandwidth B Hz. This turns out to be an important general rule, applying to digital waveforms other than the one just considered. It will be discussed in more detail in Chapter 4.

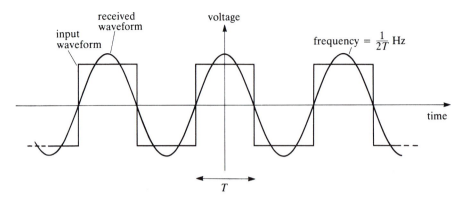

Fig. 2.11. A square wave and its fundamental Fourier component.

2.3.2   Double-sided spectra

Figs. 2.9 and 2.10 are not the only representations of line spectra. A very useful alternative method makes use of *Euler's theorem*, which states that

$$\exp(j\theta) = \cos\theta + j\sin\theta$$

Similarly

$$\exp(-j\theta) = \cos\theta - j\sin\theta$$

Adding these two expressions and rearranging leads to

$$\cos\theta = [\tfrac{1}{2}\exp(j\theta) + \exp(-j\theta)]$$

Any cosine term can therefore be re-written as the sum of two exponential terms. Consider, for example, the cosine term $\cos\omega t$, representing a sinusoidal signal with unit amplitude, angular frequency ω and zero reference phase:

$$\cdot\cos\omega t = \tfrac{1}{2}\exp(j\omega t) + \tfrac{1}{2}\exp(-j\omega t)$$

By extending our notion of the term 'spectrum', these exponential terms can also be represented on a spectral diagram, as shown in Fig. 2.12. Part (a) shows the single-sided amplitude spectrum representing $\cos\omega t$, while part (b) shows the exponential or double-sided equivalent. The two exponential terms are represented by extending the frequency axis to negative values of ω – to so-called negative frequencies.

Do not try to visualise a negative frequency – the idea of a negative frequency on its own is not helpful in this context. A real sinusoid has a frequency expressed as a positive number. If we choose to express the sinusoid as two exponential terms, then one is a function of ω and one of

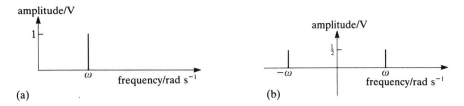

Fig. 2.12. The amplitude spectrum of the sinusoid $f(t) = \cos \omega t$: (a) single-sided; (b) double-sided.

$-\omega$. Taken together these two terms are an equivalent representation of the sinusoid. Note that the amplitude of each exponential term is half that of the original cosine term.

The notation can be extended to include phase. Consider a general sinusoidal component, and convert to exponential form:

$$A \cos (\omega t + \phi) = \tfrac{A}{2} \exp[j(\omega t + \phi)] + \tfrac{A}{2} \exp[j(-\omega t - \phi)]$$
$$= \underbrace{\tfrac{A}{2} \exp(j\phi) \exp(j\omega t)}_{} + \underbrace{\tfrac{A}{2} \exp(-j\phi) \exp(-j\omega t)}_{}$$

'positive frequency' coefficient   'negative frequency' coefficient

Coefficients have been associated with the positive and negative frequency terms. Consider first the positive frequency coefficient $\tfrac{A}{2} \exp(j\phi)$. This is a complex number in polar form with a magnitude (amplitude) $\tfrac{A}{2}$ and angle (phase) ϕ. So the complex coefficient includes both amplitude and phase information: its amplitude is $\tfrac{A}{2}$ and its phase is ϕ. Similarly the negative frequency also has an amplitude of $\tfrac{A}{2}$, but this time the phase is $-\phi$. The single- and double-sided amplitude and phase spectra of this sinusoid are compared in Fig. 2.13.

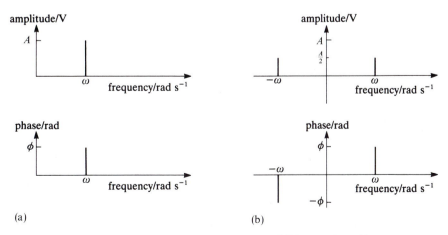

Fig. 2.13. Single- and double-sided spectra of $f(t) = A \cos (\omega t + \phi)$.

EXERCISE 2.1

Write down the exponential form, and sketch the double-sided spectra of the signals:

(a) $f(t) = 5\cos(\omega t + \frac{\pi}{2})$
(b) $g(t) = \cos \omega t + \frac{1}{2}\cos(3\omega t + \frac{\pi}{4})$

By a 'real signal' is meant a signal $f(t)$ which takes on real, rather than complex, values for all t. The mathematical theory of signal processing also deals with *complex* signals, for which the spectral symmetry just described does not hold.

It should be clear from this discussion that the double-sided amplitude spectrum of a real signal will be symmetric about the vertical axis, while the phase spectrum will be antisymmetric. This is illustrated in general terms in Fig. 2.14.

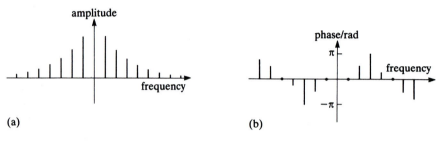

Fig. 2.14. A general line spectrum illustrating (a) the symmetric amplitude spectrum and (b) the antisymmetric phase spectrum.

The notion of a double-sided spectrum may at first seem an unnecessary complication. Its advantages should become clearer in later sections.

2.3.3 Input–output relationships

A knowledge of a system's frequency response function, either from measurement or from a mathematical model, can be used to determine the periodic steady state output of a system in response to a periodic input. Each Fourier component of the periodic input will undergo an amplitude change and phase shift as specified by the system frequency response for the particular frequency concerned. Consider an arbitrary periodic input and an arbitrary linear system with (complex) frequency response $H(f)$ – that is, with amplitude ratio $|H(f)|$ and phase characteristic $\theta(f) = \operatorname{Arg} H(f)$. In the steady state an input Fourier component at frequency f, with amplitude A_1 and phase ϕ_i will produce an output sinusoid at the same frequency but with amplitude A_o and phase ϕ_o such that

$$A_o = A_i \times |H(f)|$$

and

$$\phi_o = \phi_i + \theta(f)$$

Because of the frequency preservation property of linear systems, such a calculation can be carried out for each significant input Fourier component to obtain a complete output Fourier series. Fig. 2.15 illustrates the general case in double-sided form.

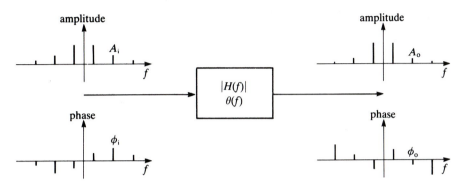

Fig. 2.15. Frequency-domain input–output relationship for a periodic signal.

As a result of the principle of superposition, the output sinusoids can be added together (taking into account their relative phase angles) to give a complete time-domain representation of the output signal. Often, however, a knowledge of the amplitude and phase of the significant transmitted components is sufficient for modelling purposes.

2.4 Frequency-domain representation of pulses

The Fourier series approach is a very useful one for periodic signals, but in general the waveforms transmitted by a digital telecommunications system will not be periodic, and will not therefore possess a Fourier series expansion. We therefore now turn to the frequency-domain representation of a single, isolated pulse. If we can model a single pulse, then by superposition we can in principle model any pulse train.

Consider first a rectangular pulse. One approach to a frequency-domain representation is to look at how the spectrum of a periodic pulse train changes as the individual pulses are separated by increasing intervals of time.

Fig. 2.16 shows a rectangular pulse train in which, unlike the waveform

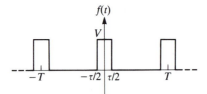

Fig. 2.16. A periodic, rectangular pulse train.

of Fig. 2.8, the high and low voltages are not of equal duration. It is no longer a square wave and has a rather different Fourier series:

$$f(t) = \frac{V\tau}{T} + \sum_{n=1}^{\infty} 2V \left[\frac{\sin (n\pi\tau/T)}{n\pi} \right] \cos n\omega t$$

where $\omega = 2\pi/T$ and T, rather than T_p, is now used to denote the repetition period. The duration of a single pulse is τ.

The single- and double-sided line spectra corresponding to this series are shown in Fig. 2.17 for the case of $\tau = T/4$. Note that:

1 the spacing between the lines is $1/T$, the repetition frequency of the pulse train;

2 the d.c. term (the mean value of $f(t)$) is equal to $V\tau/T$, in this case $V/4$;

3 the spectral lines can be viewed as possessing an envelope which becomes zero for frequencies which are an integer multiple of $1/\tau$, where τ is the pulse width;

4 the sign of groups of Fourier coefficients changes periodically as n varies. A negative coefficient is interpreted, as before, as indicating a phase angle of π or 180°.

Consider now what happens as the width τ of individual pulses is kept constant but their spacing T is increased. The spectral lines move closer

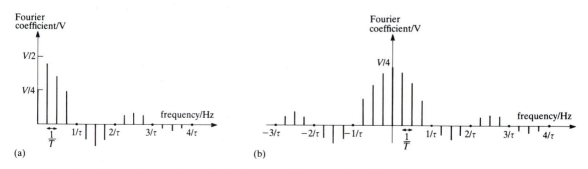

Fig. 2.17. (a) Single-sided and (b) double-sided spectra of the pulse train of Fig. 2.16.

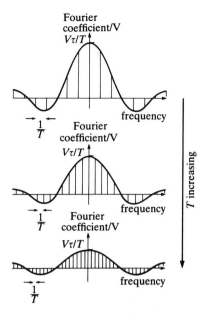

Fig. 2.18. Effect on the spectrum of increasing pulse separation T.

together and their amplitudes decrease, as shown in Fig. 2.18. As the pulse spacing is increased indefinitely, the lines become arbitrarily closely spaced, and it becomes difficult – and ultimately impossible – to distinguish between individual components. But while the amplitudes of the spectra components decrease with increasing pulse separation, the general shape of their envelope is unchanged: although the line spacing is a function of the period of the signal, the envelope (apart from its height) is determined purely by the shape of the individual pulses in the train – rectangular in this example.

This suggests that the shape of the envelope in Figure 2.18 is in some way the frequency-domain 'signature' of a rectangular pulse. This is indeed so, and the concept can be made mathematically precise by means of the Fourier transform. We shall first state the mathematical form of the Fourier transform and then try to explain its engineering significance.

2.4.1 The Fourier transform

Any practical pulse $g(t)$ can be expressed as a continuous, double-sided, distribution of spectral components $G(f)$ such that

$$G(f) = \int_{-\infty}^{+\infty} g(t) \exp(-2\pi jft)\, dt$$

There are certain mathematical conditions which must be fulfilled for $G(f)$ to exist, but these are satisfied for all practical pulses – they can be summed up by the condition that the pulse has finite energy.

If we set up and evaluate this Fourier integral for $g(t) = V$ volts for $-\tau/2 < t < +\tau/2$ and zero otherwise (in other words, a rectangular pulse of V volts and width τ seconds centred around the time origin, like the central pulse of the train in Fig. 2.16) we obtain the expression

$$G(f) = V\tau\left(\frac{\sin \pi\tau f}{\pi\tau f}\right)$$

The derivation of this expression is given, along with some other properties of the Fourier transform, in Appendix A.

It has been written in the way shown so that the expression inside brackets takes on the form $(\sin x)/x$, which offers a certain mathematical convenience. Its graph is plotted in Fig. 2.19: it is the function describing the envelope of the line spectrum of the rectangular pulse train and can be interpreted as the spectrum of a single isolated rectangular pulse centred around the time origin.

Note that $G(f)$ does not possess any spectral lines: only signals with periodic components possess spectra which include discrete lines. So no individual frequency components can be identified in the spectrum of an isolated pulse. This should seem reasonable from the previous discussion of the rectangular pulse train. The spectral lines were a consequence of periodicity, and their spacing was determined by the pulse spacing, or period, of the signal. For an isolated pulse, the term pulse spacing or period has no meaning: no individual lines can be identified in the spectrum, which is a continuous distribution of spectral components. Only the shape of the individual pulse is reflected in the form of $G(f)$.

The arbitrarily closely spaced spectral lines of the train of widely separated pulses have become the continuous spectral distribution of the isolated pulse.

In what sense, then, is $G(f)$ a spectrum? Consider first its dimensions. Looking back at the mathematical expression, we see that $\pi\tau f$ is dimensionless, so the dimensions of $G(f)$ are [volts] × [time]. The area under the graph of $G(f)$ v. f therefore has the dimensions of voltage. Because of this, $G(f)$ is often expressed as 'volts per hertz' (V Hz^{-1}), and is referred to as the spectral density. As with other density functions, it is the

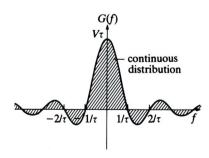

Fig. 2.19. A continuous distribution of spectral components.

area under a spectral density function which can be related to a physical quantity. The area under the curve between two frequencies can be thought of as a measure of the contribution of that particular *range* of frequencies to the overall pulse. Note how this interpretation contrasts with that of a spectral line, whose *height* represents the amplitude of a discrete sinusoidal component at a *single* frequency.

For a periodic signal, a spectral line at zero frequency represents the d.c. term or mean value A_0. The value of a continuous spectrum at zero frequency, $G(0)$, also has a useful physical interpretation. From the form of the Fourier integral it follows that

$$G(0) = \int_{-\infty}^{+\infty} g(t)\,dt$$

This is equal to the area of the pulse: $V\tau$ in the case of a rectangular pulse. Other features of $G(f)$ for any particular pulse will depend on its precise shape. Note that the zero-crossings of the spectrum of the rectangular pulse take place at integral multiples of $1/\tau$, where τ is the pulse width.

The Fourier transform $G(f)$ of a pulse $g(t)$ is a unique frequency-domain representation of the pulse, just as a Fourier series is a unique representation of a periodic signal. In general $G(f)$ takes on a complex value for each value of frequency. In such cases, $|G(f)|$ represents a continuous distribution of amplitudes, while Arg $G(f) = \phi(f)$ is a continuous phase spectrum. For an even pulse where $g(t) = g(-t)$, such as the rectangular pulse drawn symmetrically about the origin, $G(f)$ is real, and can be represented on a single spectral diagram. A negative value of $G(f)$ for a particular frequency – rather like a negative Fourier series coefficient – means that $\phi(f)$ has a value of $\pm\pi$ (or an odd multiple thereof) for that frequency. Almost all the pulses discussed in this book will be even, with real spectral densities $G(f)$.

Thinking too hard about the physical meaning of continuous distributions of sinusoidal components can sometimes be counterproductive, and it is often better to view $G(f)$ simply as an alternative and equivalent mathematical representation of a pulse $g(t)$. The $g(t)$ form is the time-domain representation, while $G(f)$ is the equivalent frequency-domain model. The important thing is to be able to move easily between the two domains when modelling a telecommunications system.

The engineering significance of such highly abstract and idealised models will be discussed below.

EXERCISE 2.2

Sketch the spectrum $G(f)$ of each of the pulses shown in Fig. 2.20. Label the numerical values of the zero crossings and the height of the spectrum at $f=0$.

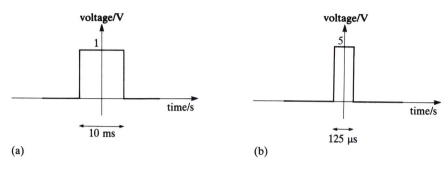

Fig. 2.20. Pulses for Exercise 2.2.

EXERCISE 2.3

Fig. 2.21 shows the spectra of two rectangular pulses. Sketch each pulse, clearly labelling their height and width.

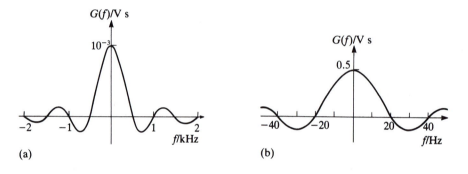

Fig. 2.21. Rectangular pulse spectra for Exercise 2.3.

2.4.2   Standard models and general 'rules'

The spectrum of the rectangular pulse illustrated in Fig. 2.19 is such an important theoretical tool that it takes first place in a 'library' of standard models.

A second standard model can be derived from the rectangular pulse spectrum by considering what happens to the amplitude spectrum $V\tau\,|\,(\sin \pi\tau f)/\pi\tau f\,|$ as the rectangular pulse is decreased in width, keeping its area or 'strength' constant. Fig. 2.22 illustrates this process. Note that as the pulse becomes narrower, its spectrum becomes broader, with the side lobes remaining significant at higher and higher frequencies. This is an

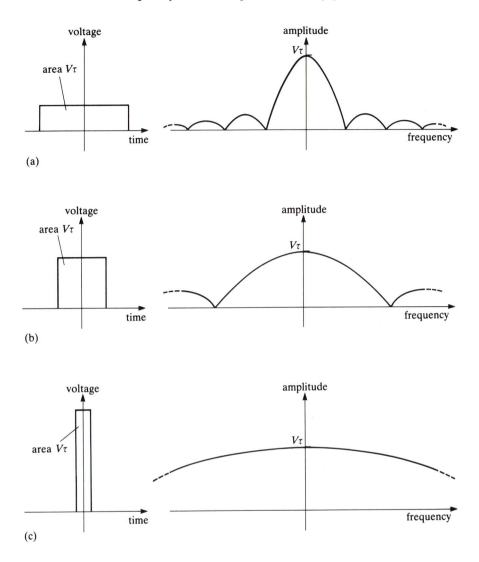

Fig. 2.22. Effect on the amplitude spectrum of decreasing the width of a rectangular pulse.

important general rule, and is worth highlighting as the first of a number of general properties of time- and frequency-domain models:

The shorter the pulse, the broader its spectrum.

Sufficiently short pulses can be treated as though they possess a flat spectrum over an indefinitely wide range of frequencies. The expression 'sufficiently short' means that the duration of the pulse is much shorter than the response time of any part of the system under consideration. Such a pulse can be characterised completely by its 'strength' or area A; its precise shape is immaterial. The mathematical model of such an indefinitely narrow pulse is known as an *impulse* or *Dirac delta function* $\delta(t)$; it is represented diagrammatically by a vertical arrow at the time origin. The spectrum of a delta function of strength A volt seconds is a constant spectral density of A volts per hertz, as can be deduced from Fig. 2.22 and illustrated specifically in Fig. 2.23.

Two more standard pulse models are shown (together with the rectangular pulse for comparison) in Fig. 2.24. All the pulses are drawn with equal areas and height, so it is easier to compare their spectra. This figure illustrates another important general rule about time- and frequency-domain models. Compare, for example, the spectra of the rectangular and triangular pulses. The side-lobes of the latter decrease with increasing frequency much faster than do those of the former. In fact, the spectrum of the triangular pulse is given by the expression

$$V\tau \frac{\sin^2 \pi\tau f}{(\pi\tau f)^2}$$

compared with

$$V\tau \frac{\sin \pi\tau f}{\pi\tau f}$$

for the rectangular pulse. Hence the heights of the side-lobes of the rectangular pulse spectrum are inversely proportional to frequency, whereas those of the triangular pulse are inversely proportional to the *square* of the frequency. The effect is even more pronounced for the third, more rounded, pulse illustrated, which is a 'raised-cosine' pulse, consisting of one cycle of a cosine wave raised by adding a d.c. level equal to its amplitude: in this case the side-lobes of the spectrum of the raised-cosine pulse rapidly become negligible. In fact, their heights are inversely proportional to the *cube* of the frequency (although the complete mathematical expression for the spectrum has a rather more complicated form than simply $(\sin x/x)^3$).

Now compare the time-domain characteristics of the pulses. The rectangular pulse has two jump discontinuities: sudden changes in voltage at the beginning and end of the pulse. The triangular pulse has no sudden jumps in voltage level, but its slope changes suddenly three times. In the case of the raised-cosine even the slope changes smoothly over the whole time interval. This leads to the second general 'rule':

Sudden changes in a pulse shape imply high frequencies in the corresponding spectrum.

The final models in our library follow immediately from the preceding ones as a result of an important general property of the Fourier transform.

In practical terms we can say that the spectrum is constant for all frequencies to which any element of the system under consideration can respond; such an 'indefinitely wide' spectrum is the frequency-domain equivalent of a 'sufficiently short' pulse in the time-domain.

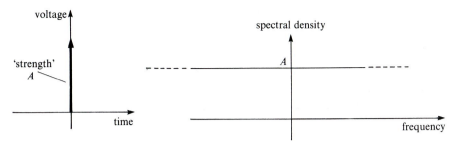

Fig. 2.23. Spectrum of $\delta(t)$, an ideal impulse occurring at time $t=0$.

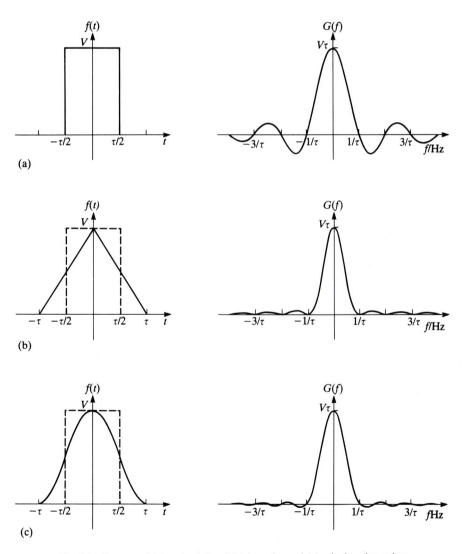

Fig. 2.24. Spectra of (a) rectangular, (b) triangular and (c) raised-cosine pulses.

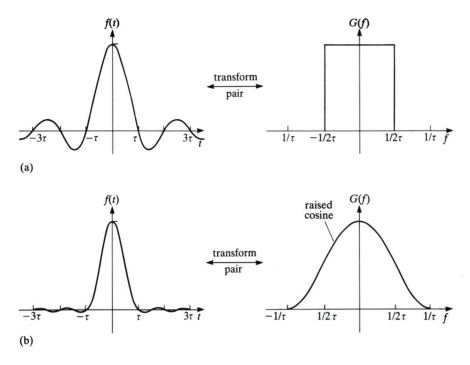

Fig. 2.25. Pulses with (a) rectangular and (b) raised-cosine spectra.

As already mentioned, the relationship between a pulse and its spectrum is unique. The mathematical functions describing the two are said to form a Fourier transform pair: Fig. 2.24 shows three such transform pairs. One interesting characteristic of a Fourier transform pair is that for an even signal, for which $f(t)=f(-t)$, the time- and frequency-domain functions can be interchanged, as illustrated in Fig. 2.25. By this is meant that the time-domain 'pulse' corresponding to a perfectly rectangular spectrum has exactly the same $(\sin x)/x$ shape as the spectrum of a rectangular pulse. Similarly, the pulse corresponding to a raised-cosine spectrum has the form shown in Fig. 2.25(b). Note that this perfect symmetry between time- and frequency-domains is only possible when the double-sided form of a spectrum is employed.

Both of the time-domain waveforms of Fig. 2.25 (and the corresponding one for a triangular spectrum) extend, in theory, over all time. Any physically realisable waveform, however, must have at least a defined beginning! This does not detract from the usefulness of such models in practical engineering, however, since they lead to important general conclusions of the type already stated, as well as enabling us to model various worst-case situations.

One more such general result follows directly from Fig. 2.25. Note that the time-domain waveform corresponding to the rectangular spectrum is much more oscillatory than that corresponding to the raised-cosine spectrum; the pulse with a triangular spectrum would be somewhere in between. In fact, we have a third 'rule':

Sudden changes in a spectrum imply oscillatory behaviour in the corresponding signal.

2.4.3   The engineering interpretation

All the models introduced so far are mathematical abstractions. Nevertheless, they are all of profound significance for engineering. One of their most important uses is as idealisations of particular practical features or limitations of telecommunications systems. For example, if we wish to transmit data at very high rates, then we need to use very short pulses and the first of the above general rules implies that high bandwidths will be required. Similarly, a consequence of the second rule is that in order to restrict the spectrum of a pulse to a particular bandwidth we need to tailor its shape to avoid sharp transitions. And the third 'rule' tells us that the sharper the spectral cut-off of a filter used in a telecommunications system, the more oscillation or 'ringing' will be observed in the time-domain.

These three examples illustrate typical compromises which have to be made: we might wish to transmit data at very high rates, but only have a limited bandwidth at our disposal; rectangular pulses might be easiest to generate, but their strong high frequency components due to the sudden time-domain transitions may be a problem; sharp filter cut-offs might well be desirable to maximise the use of available bandwidth, but the 'ringing' of time-domain signals may be highly *undesirable*. The mathematical descriptions of idealised pulses and their spectra provide a modelling framework within which to evaluate such trade-offs and compromises.

2.4.4   Input–output relationships

Input–output relationships in the frequency domain can be obtained for pulses in a way completely analogous to the procedure for line spectra. The only difference is that continuous distributions of frequency components are involved, rather than individual spectral lines. Suppose that the input pulse to a linear system with frequency response $H(f)$ has an amplitude spectrum $|G_i(f)|$ and phase spectrum $\phi_i(f) = \text{Arg } G_i(f)$. Then

$$|G_o(f)| = |G_i(f)| \times |H(f)|$$

and

$$\phi_o(f) = \phi_i(f) + \theta(f)$$

where $\theta(f)$, the phase characteristic of the system, is equal to $\text{Arg}\, H(f)$ as before.

As with the periodic signals, we *multiply* the amplitude functions and *add* the phase terms. These amplitude and phase input–output relationships can be written even more succinctly in terms of the *complex* spectra $G_i(f)$ and $G_o(f)$, and the *complex* frequency response function $H(f)$. All three expressions contain the required amplitude and phase information and we have:

$$G_o(f) = G_i(f) \times H(f)$$

Like pulse spectra, a frequency response function $H(f)$ can be used in single- or double-sided form. Unlike pulse spectra, however, the numerical values of $|H(f)|$ are identical in both forms, since they represent amplitude *ratios* applied equally to sinusoidal (single-sided) or exponential (double-sided) components.

2.5 Characterising systems in the time-domain

The notion of equivalent models in time- and frequency-domains has already been stressed. So, for example, the time-domain function $g(t)$ modelling a *signal* is a completely equivalent model to the frequency spectrum $G(f)$. They form a Fourier transform pair, often written

$$g(t) \leftrightarrow G(f)$$

(Note the convention of using of a lower case g for the time-domain and upper case G for frequency-domain.)

In the same way there is a time-domain characterisation of a linear *system* which turns out to be completely equivalent to its frequency response function $H(f)$. The key to this representation is the way the system responds to an input pulse of duration very much shorter than its response time.

2.5.1 Impulse response

Fig. 2.26 shows the response of a first-order lowpass system with unity d.c. gain to an input rectangular pulse. In part (a) the duration of the pulse is equal to the system time constant, and the pulse response exhibits the characteristic rise to 63% of the input value, followed by exponential decay.

The rising portion of the pulse response is part of the system unit step response $1 - \exp(-t/\tau)$, where $\tau = 1/\omega_c$ is the system time constant and ω_c is the angular 3 dB cutoff frequency.

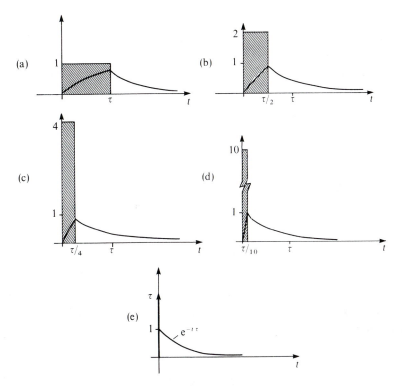

Fig. 2.26. Impulse response of a first-order lowpass system as the limiting case of narrow rectangular pulse response. The system has unity low frequency gain, and all input pulses have area τ.

The other parts of the figure illustrate what happens as the width of the pulse is decreased, keeping its area or strength constant at τ units. In the limit of a very short input pulse, the system response is independent of the precise pulse shape or duration and is characteristic of the system alone: only the strength of the impulse affects the output, scaling it linearly. Such an indefinitely short pulse was introduced earlier as one of our standard models – it is the delta function or impulse $\delta(t)$. The response to an impulse of unit strength – the *unit impulse response* – is therefore often used to characterise a system. Fig. 2.26 shows how the response of a first-order lowpass system with unity d.c. gain converges to $\exp(-t/\tau)$ as the input becomes closer and closer to an impulse of area τ units. The unit impulse response of a general first-order lowpass system with time constant τ and d.c. gain k is similarly $\frac{k}{\tau}\exp(-t/\tau)$.

The derivation of such expressions is not important in the present context, but can be found in Meade & Dillon.

2.5.2　Relationship to frequency response

One of the advantages of the unit impulse response is that it is very closely related to the frequency response function of a linear system. From

Section 2.4.2 the spectrum of an impulse of strength A Vs is simply A VHz^{-1}. So the spectrum of a unit impulse is a constant unit spectral density. Using the input–output relationship in the frequency domain we therefore have

$$\text{output spectrum} = \text{input spectrum} \times \text{frequency response}$$

or, in this case

$$\text{impulse response spectrum} = 1 \times H(f).$$

The spectrum of the impulse response of a linear system is therefore identical to the system's frequency response. Impulse response is usually given the symbol $h(t)$ and forms a Fourier transform pair with the frequency response:

$$h(t) \leftrightarrow H(f).$$

A common way of modelling a transmitted pulse in a digital telecommunications system is to imagine the pulse being generated by applying an impulse to a suitable linear system or 'transmit filter'. Then $g(t)$, the pulse profile, is viewed as the impulse response of the transmit filter whose frequency response $G(f)$ is identical to the pulse spectrum. This model does not correspond exactly to what is done in practice, but it is a useful idealisation which can be used to compare different pulse generation and coding systems. This will be discussed in more detail in Part 2.

Note also the implications of this result for the realisability of an ideal, distortionless 'brick-wall' filter with the frequency response $H(f)$ illustrated, in double-sided form in Fig. 2.27. The impulse response $h(t)$ of such a filter would be of $(\sin x)/x$ form, and would need to be symmetrical about time $t=0$, beginning well before the impulse is applied to the system! Such a *non-causal* (and thus non-realisable) filter or channel corresponds to part (a) of Fig. 2.5, where not even a constant propagation

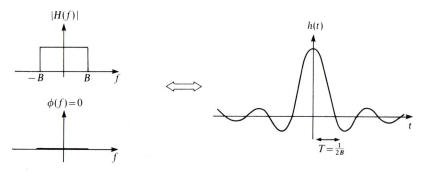

Fig. 2.27. Frequency response and impulse response of a brick-wall filter (or distortionless channel) with zero time delay.

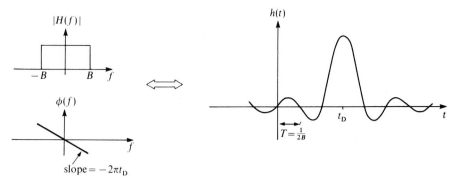

Fig. 2.28. Frequency response and impulse response of a brick-wall filter or (distortionless channel) with transmission delay t_D.

delay for all frequencies was permitted. Allowing a constant delay for all frequencies, as in Fig. 2.5(b), leads to the frequency response model of Fig. 2.28, with an amplitude ratio $H(f)$ which is constant over the filter bandwidth as before, but now with a linear phase characteristic $\phi(f)$ instead of constant *zero* phase as in Fig. 2.27. The new phase characteristic has the result of delaying the impulse response by the propagation delay t_D. The system is still theoretically unrealisable, however: the tails of a perfect $(\sin x)/x$ function still extend over all time, whereas the impulse response of any practical system can only begin after the application of the impulse. Within certain limitations, *approximations* to the brick-wall response can be realised by allowing a suitable delay and truncating the impulse response so that it can begin at time $t = 0$. Even more importantly, however, the idealised model remains a theoretical tool highly useful in its own right; it will recur at intervals throughout this book.

2.6 Discrete processing

At various stages in digital telecommunication systems analogue signals are sampled and processed in discrete form. The analogue signal concerned could be a speech signal from a telephone transmitter, or it could be a pulse distorted and contaminated by noise after transmission over a non-ideal telecommunications channel. Telecommunications engineers therefore need to be familiar with some of the principles of discrete signal processing, although the topic will be treated here very briefly.

Fig. 2.29 shows the major stages in the discrete processing of an analogue signal. The value of the analogue signal is first sampled, almost always at regular intervals of time. These sample values are then often (but

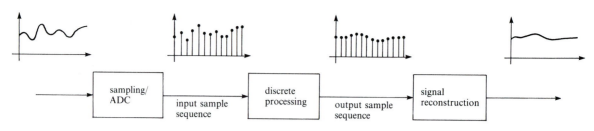

Fig. 2.29. Discrete processing of an analogue signal.

not always) converted into a digital representation (digitised) by an analogue to digital converter (ADC). The discrete (analogue or digital) sample values are then processed by electronic hardware or software, producing an output in the form of another sequence of samples. Finally, this output sequence will normally need to be converted back into the form of a continuous signal by means of a digital to analogue converter (DAC) and/or suitable filtering – a process known as signal reconstruction. In this section we shall briefly consider sampling, discrete processing and signal reconstruction. Digitisation will be considered further in Chapter 7.

> Switched-capacitor filters, widely used in signal processing for telecommunications, are examples of *discrete* but not *digital* processing. In such devices, the sampled signal value is represented by the charge on a capacitor, so that no conversion to digital form is required. Charge is switched from capacitor to capacitor at the sampling instants in order to carry out the discrete processing operations.

2.6.1 Sampling

Fig. 2.30 illustrates a vitally important feature of the sampling process. In part (a) the signal changes smoothly, without fluctuation, between the sampling instants. As long as this is the case – and the proviso will be quantified below – smooth interpolation between the sample values will recover the message signal perfectly. In part (b) on the other hand, fluctuations occur between the sampling instants and these are lost on reconstruction by smooth interpolation. Finally, part (c) shows an example where the interpolated waveform bears very little resemblance to the message signal: it is a completely different sinusoid. The general phenomenon is known as *aliasing*, and the improperly reconstructed sinusoid of part (c) is known as a low-frequency alias of the original signal.

To reconstruct a sampled signal without aliasing, the sampling rate must be matched to the frequency range of the original signal. The rule governing the relationship is generally attributed to Nyquist and/or Shannon, and is known as the sampling theorem. A statement of the sampling theorem for baseband signals is:

If the frequency components present in a continuous signal extend from 0 to B Hz, then the signal can be completely represented by, and reconstructed

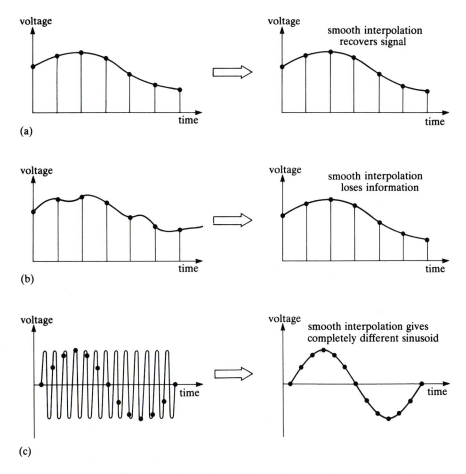

Fig. 2.30. The effect of sampling and smooth interpolation on three message signals.

from, a sequence of equally spaced samples, providing that the sampling frequency exceeds 2B samples per second.

The sampling theorem also exists in more general forms covering bandlimited signals which do not extend to d.c. and also to signals which are sampled at irregular intervals. Such matters are beyond the scope of the present text.

Parts (b) and (c) of Fig. 2.30 show situations where the sampling theorem is violated and aliasing occurs. In both cases the signal contains components above half the sampling frequency. Information about the signal is lost as a result of such improper sampling and reconstruction.

Just as both periodic signals and aperiodic pulses can be uniquely represented by their spectra, so a sampled signal has a corresponding frequency-domain representation. In this case, the frequency-domain representation includes a feature which models the potential for aliasing.

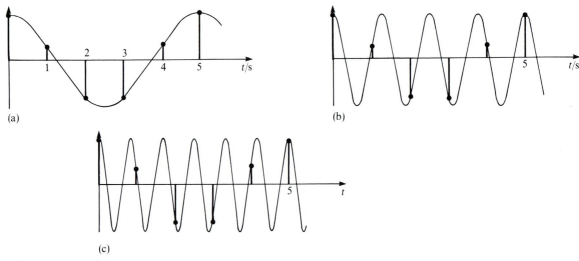

(a)

(b)

(c)

Fig. 2.31. The same sequence of samples generated by sinusoids of three different frequencies.

Fig. 2.31(a) shows a sinusoid of frequency $\frac{1}{5}$ Hz (period 5 s), sampled at a rate of 1 Hz – that is, well in accordance with the sampling theorem. But these samples could also have been generated by a sinusoid with a frequency $\frac{4}{5}$ (0.8) Hz or $\frac{6}{5}$ (1.2) Hz (again sampled at 1 Hz) as shown in parts (b) and (c) of the figure. Sinusoids of frequency $\frac{9}{5}$ or $\frac{11}{5}$ Hz, $\frac{14}{5}$ or $\frac{16}{5}$ Hz, and so on would have the same effect. In fact, the samples have the potential to represent an infinite number of aliased sinusoids. This potential is reflected in the corresponding spectrum, shown in Fig. 2.32. It includes not only the spectral lines of the original signal at $\pm\frac{1}{5}$ Hz, but also an infinite set of equally spaced replicas of these lines centred around multiples of the sampling frequency. Exactly the same applies to a signal with a continuous, rather than a line spectrum: the spectrum of the sampled version consists of equally spaced replicas of the original

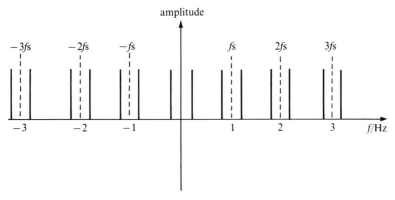

Fig. 2.32. Amplitude spectrum of a sampled sinusoid.

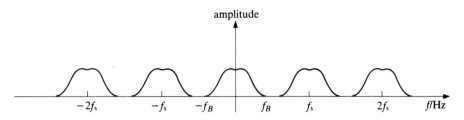

Fig. 2.33. Amplitude spectrum of a general sampled signal.

spectrum. Fig. 2.33 shows an example where the original signal has a continuous amplitude spectrum extending from 0 to f_B and the condition $f_B < f_s/2$ holds in accordance with the sampling theorem. It can also be shown that a sampled signal has a similarly repeating phase spectrum.

Note that the amplitude axis has not been labelled with numerical values in Figs. 2.32 and 2.33. To quantify the height of the sampled spectra we need a rather more formal model of a 'sample'. Fig. 2.34(a) shows a common representation of the sampling process as a switch. The switch is assumed to close (terminal A) for a brief period every T seconds, where $T = 1/f_s$ is the sampling interval, thus sampling the value of the input $x(t)$ at regular intervals. The output of the switch is illustrated in Fig. 2.34(b). It can be shown that, provided the pulses representing the sample values are short enought to be treated as ideal impulses, then the amplitude spectrum of the pulse waveform $y(t)$ approximates very closely the form of Fig. 2.33 – that is, a periodically repeating version of the amplitude spectrum of $x(t)$. If each sample is represented by an impulse numerically equal in strength to the sample value itself, then the height of the repeating amplitude spectrum turns out to be $1/T$ times the height of the message spectrum. Telecommunications engineers would not often need to be concerned with the numerical values of such amplitudes, and readers interested in the mathematical details are referred to Peebles, Lynn & Fuerst, or Meade & Dillon. The general, repeating, form of the spectrum of a sampled signal is, however, of great practical significance.

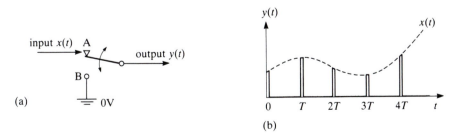

Fig. 2.34. A switch model of the sampling process: (a) the sampler; (b) the input signal $x(t)$ and the sampled output $y(t)$.

2.6.2   Signal reconstruction

The frequency-domain representation of a sampled signal is a very useful model, since it illustrates in theory how a signal can be reconstructed from its samples, and it also provides a context for interpreting the sampling theorem. Consider again Fig. 2.33, and assume it is the spectrum of a signal in which each sample value is represented by an impulse of appropriate strength. At the low-frequency end is the spectrum of the original message signal (apart from a scaling depending on the impulse strength). The message signal can therefore be recovered simply by passing the impulse waveform through a lowpass filter which passes all components up to f_B unchanged, and completely removes the high-frequency replicas.

Suppose now that the sampling frequency is (a) increased and (b) decreased. The effect on the sampled spectrum is shown in Fig. 2.35. Increasing the sampling frequency as in part (a) simply increases the spacing between the replicas of the message signal; lowpass filtering will still recover the original. In part (b) the sampling frequency has been reduced such that $f_B > f_s/2$ in violation of the sampling theorem. The repeated spectra overlap, and the output of a lowpass filter would not be a faithful replica of the message signal.

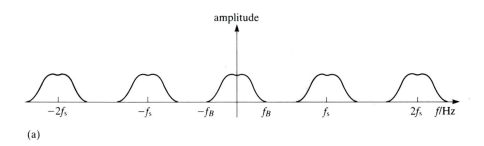

(a)

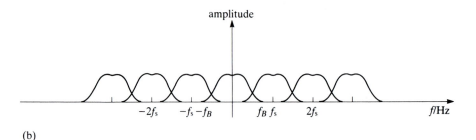

(b)

Fig. 2.35. Effect in the frequency domain of (a) increasing and (b) decreasing the sampling rate in comparison with Fig. 2.33.

The precise values of sampling frequency employed, and the details of the reconstruction filter, will depend on the particular system. Some information about the system used for voice telephony will be given in Chapter 7.

The minimum sampling frequency, as specified by the sampling theorem, is just sufficient to prevent the replicas from overlapping. An ideal lowpass filter, with infinitely sharp cut-off and linear phase, would then recover the original. In practice, of course, sampling rates are chosen to be higher – often much higher – than the theoretical minimum, so that a practical, realisable filter can be used for signal reconstruction without introducing too much amplitude or phase distortion.

2.6.3 Characterising discrete systems

A discrete linear system is one which produces an output sample sequence $y(n)$ in response to an input sequence $x(n)$, obeying the principle of superposition. (A version of the frequency preservation property also holds, although a little care is needed to take into account the aliasing phenomenon and the infinitely repeating nature of the spectra of sampled signals.) As with continuous systems, there are two equivalent ways of characterising discrete linear systems – one in the frequency-domain and one in the time-domain. The former will not be considered here in any detail. Suffice it to say that a frequency response function can be defined which specifies, for all frequencies, the amplitude ratio and phase shift introduced by the system in the steady state when an input sampled sinusoid gives rise to an output sampled sinusoid of the same frequency. This is illustrated in Fig. 2.36. For further information the reader is referred to Lynn & Fuerst, Meade & Dillon, and Bissell.

The time-domain characterisation is of more interest here, however, and is particularly straightforward. Just as the unit impulse response completely specifies a *continuous* linear system, so the unit sample response completely specifies a *discrete* linear system. Fig. 2.37 shows two

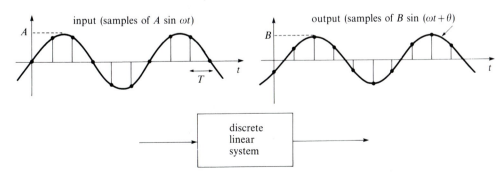

Fig. 2.36. 'Frequency preservation' property of a discrete linear system.

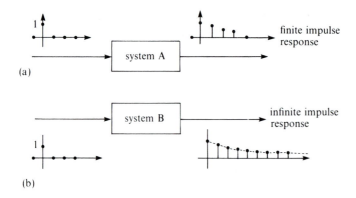

(a)

(b)

Fig. 2.37. Discrete linear systems with finite and infinite unit sample response sequences.

contrasting unit sample response sequences. In part (a) the unit sample response is a finite length sequence, whereas in part (b) the response is (at least in theory) an infinite sequence whose sample values decay exponentially, never reaching zero. These two categories are usually referred to as *finite impulse response* (*FIR*) and *infinite impulse response* (*IIR*) systems respectively. Discrete processing of both types is widely used in digital telecommunications.

Fig. 2.38 is a block diagram of a *transversal filter*, a simple type of FIR processor. The blocks labelled T are used to store or delay the value of a sample for the duration of one sampling interval, before releasing it to the next stage. The sample values and their delayed counterparts are then

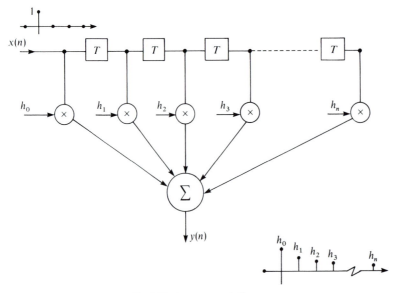

Fig. 2.38. A transversal filter.

used to generate a weighted sum, which ultimately forms the output of the filter. Transversal filters can be implemented in a variety of ways using tapped delay lines, SAW (surface acoustic wave) devices, or digital hard- or software. The unit sample response of a transversal filter is closely related to the coefficients of the multiplier blocks. It follows directly from the figure that the response to a single input unit sample is a sequence of values determined directly by the successive multiplier coefficients:

$$h(n) = h_0, h_1, h_2, h_3, \ldots . h_n$$

Some uses of such filters in digital telecommunications will be considered later. For the theoretical background see Lynn & Fuerst.

2.7   Input–output relationships in the time-domain

To conclude this chapter we present a new type of input–output relationship for linear systems: one based on time-domain rather than frequency-domain models. The theoretical development is easier to introduce in the context of discrete signals and systems.

2.7.1   Discrete systems

Because of the principle of superposition, the output of a discrete linear processor to any input sequence can easily be determined by superposing appropriately weighted unit sample response sequences. The process is illustrated in Fig. 2.39. The system's (finite) unit sample response is shown at the top of the figure, and below it is the input sequence whose corresponding output sequence we wish to calculate. The next four lines show how the response to each of the four samples of the input sequence can be calculated individually. Because of linearity and the principle of superposition these individual responses can be summed to give the complete system output, shown in the bottom line of the figure. Note how the output response sequences is longer than the input. In this example the input sequence extends over a period of time equal to 3 sample intervals, and the unit sample response sequence over 2 sample intervals. As a result, the output sequence extends over $3 + 2 = 5$ sample intervals.

A spreadsheet is ideal for carrying out numerical calculations like this. Some applications of spreadsheets for modelling in telecommunications are described in Appendix C.

A mathematical expression for this process, known as convolution, follows immediately from the figure. Denoting the system's unit sample response by

$$h(n) = h_0, h_1, h_2, \ldots$$

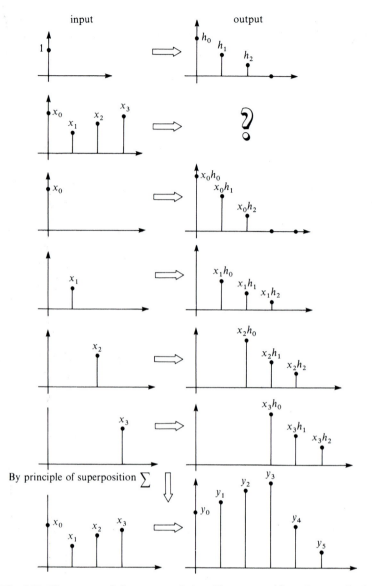

Fig. 2.39. The process of discrete convolution. The superposition of appropriately weighted and delayed unit sample response sequences gives the response of a discrete linear processor to an arbitrary input sequence.

we have, from the figure,

$$y_0 = x_0 h_0$$
$$y_1 = x_0 h_1 + x_1 h_0$$
$$y_2 = x_0 h_2 + x_1 h_1 + x_2 h_0$$

and so on.

Note the pattern. Each output sample y_n is the 'convolution sum' of a number of products of terms from the input and unit sample response sequences. The subscript of the input terms *increases* from left to right, while that of the unit sample response terms *decreases*. For a general output term y_i of any discrete linear system we can write

$$y_i = x_0 h_i + x_1 h_{i-1} + x_2 h_{i-2} + \ldots x_{i-1} h_1 + x_i h_0$$

or, using compact summation notation,

$$y_i = \sum_{k=0}^{i} x_k h_{i-k}$$

The convolution process is often written directly in terms of input, output and impulse response sequences as

$$y(n) = x(n) * h(n)$$

where the symbol * denotes the convolution operation carried out as just described for all sample values.

An equivalent, alternative form of the convolution expression is obtained by modifying the order in which the terms are summed:

$$y_i = \sum_{k=0}^{i} h_k x_{i-k}$$

EXAMPLE

A discrete processor has the unit sample response sequence

$$2, 1.5, 1.0, 0.5, 0, 0, 0, \ldots$$

The sample sequence

$$1, 2, 3, 4, 5, 6, \ldots$$

is applied to the processor input. What is the fourth output sample (y_3)?

SOLUTION

From the general convolution expression we have

$$\begin{aligned}
y_3 &= x_0 h_3 + x_1 h_2 + x_2 h_1 + x_3 h_0 \\
&= 1 \times 0.5 + 2 \times 1.0 + 3 \times 1.5 + 4 \times 2 \\
&= 15
\end{aligned}$$

2.7.2 Continuous systems

An analogous procedure can be used for continuous linear systems. The basic idea is to view the input signal as consisting of contiguous short pulses. In the limit of an 'indefinitely large' number of 'indefinitely short' such pulses, each can be treated as an impulse. The total system output can then be calculated by superposing the individual impulse responses. Fig. 2.40 gives the general idea. Suppose for simplicity that the impulse response of the system is a rectangular output pulse as shown at the top of

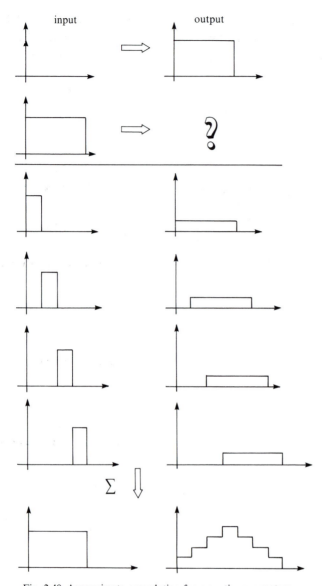

Fig. 2.40. Approximate convolution for a continuous system.

the figure, and we wish to calculate the system response to an input rectangular pulse, as shown in the second line. Proceeding in a way completely analogous to the discrete system above, we decompose the input signal into short pulses. Assuming that these pulses are short enough to be treated as impulses, then the individual system impulse responses can be summed to find the total response. (Note that we have not attempted to quantify the strength of the individual short pulses.) The total system output therefore has the general form shown as the last line of the figure.

Now, unlike Fig. 2.39, which is a *precise* description of discrete convolution, Fig. 2.40 is merely an *approximation* to the continuous case. It shows an input signal decomposed into just four short pulses! Yet to use system impulse response as a proper basis for an input–output relationship, the input signal must be viewed as an indefinitely large number of contiguous impulses or delta functions. The usual mathematical technique for doing this is to set up a series for finite width pulses, very similar in form to the summation expression above, and then convert it to an integral expression by letting the individual pulses tend to zero width. The procedure is not difficult, and details are given in Appendix B. Here we simply quote the result and explain its use.

The output $y(t)$ of a linear system is given by the convolution of the input $x(t)$ with the system impulse response $h(t)$. This is usually written $y(t) = x(t)*h(t)$, where

$$x(t)*h(t) = \int_{-\infty}^{+\infty} x(z)h(t-z)\,dz$$

The variable z is a 'dummy' variable. In other words, the integration is performed with respect to z, leading initially to an expression containing both z and t. Once the limits of integration are substituted however, the variable z disappears, leaving a final expression for y as a function of t, as desired. Engineers do not often need to evaluate such expressions, but a 'feel' for the effect of convolution is a useful practical skill, and one which can best be acquired by means of the following graphical interpretation of the convolution integral.

From the general form of the integral it is clear that to obtain the value of the output signal $y(t)$ at time t, we form the function $x(z) \times h(t-z)$ and integrate. In other words, find the area under the curve representing $x(z) \times h(t-z)$ *plotted against z*. The function $x(z)$ plotted against z has exactly the same shape as $x(t)$ against t: it is the input signal profile. Fig. 2.41 illustrates the significance of $h(t-z)$ for $h(t) = \exp(-t)[t>0]$, the impulse response of a first-order lowpass system with unity gain and time constant. The function $h(-z)$ plotted against z is simply the reversed profile of the impulse response, and is shown in part (c) of the figure. The function $h(t-z)$ is the same profile shifted by a distance t, as shown in part (e). The convolution integral can therefore be interpreted graphically as follows: the area beneath the graph of $x(z)$ multiplied by $h(t-z)$ for any time shift t represents the system output $y(t)$ at time t.

That this is all not quite so daunting as it appears should be illustrated by the following examples!

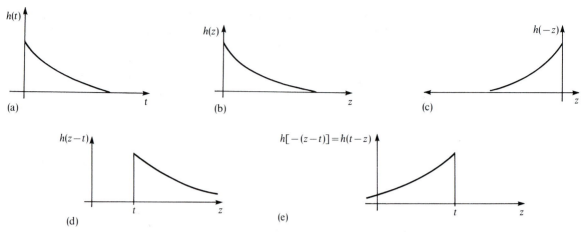

Fig. 2.41. Setting up the functions for the convolution integral. (Note that (e) can be
viewed as either a time-shifted (c) or a time-reversed (d).

EXAMPLE 1: THE UNIT STEP RESPONSE OF A
FIRST-ORDER LOWPASS SYSTEM

This is illustrated in Fig. 2.42. Part (a) shows the impulse response of
the system, and part (b) the unit step input. The reversed and shifted
impulse response $h(t-z)$ is shown overlapping the step input, now
$x(z)$, in part (c). Because this $x(z)$ is unity for all $z > 0$ and zero
otherwise, the integral of the function $x(z) \times h(t-z)$ with respect to z

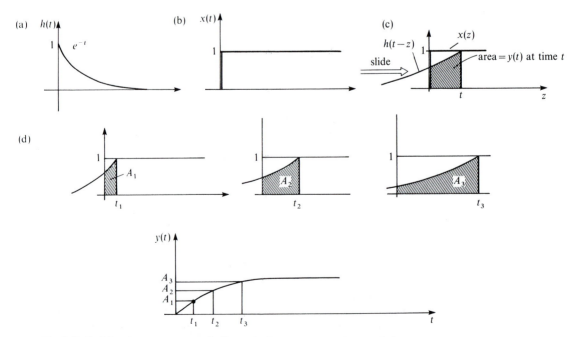

Fig. 2.42. Deriving the step response of a first-order lowpass system via convolution.

for time shift t is simply the shaded area shown. Outside this area one or other of the overlapping functions has zero magnitude, and the product is therefore also zero.

The complete step response is given by calculating the shaded area for all time shifts t from 0 to ∞. Three such areas are illustrated in part (d) of the figure, showing how they relate to the system step response $y(t)$ at the three corresponding values of t.

EXAMPLE 2: THE CONVOLUTION OF TWO RECTANGULAR PULSES

Fig. 2.43 shows the application of the convolution integral to the situation modelled approximately in Fig. 2.40. Instead of dividing up the input into a small number of contiguous pulses, we apply the graphical method which accurately reflects the integration. Proceeding as in Example 1, we slide the reversed impulse response over the rectangular pulse input. Multiplying and calculating the area of overlap leads to the triangular output pulse in part (d) of the figure. Note that because the impulse response is symmetric the procedure becomes rather easier. Note also that the system output has a total duration equal to the sum of the durations of the input and the impulse response. This corresponds to earlier experience in the discrete case, and is a general rule where both input and system impulse response are of finite duration.

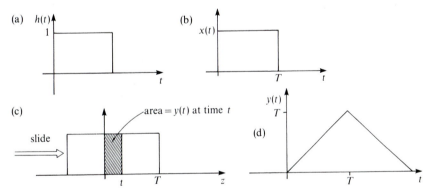

Fig. 2.43. Convolution of two rectangular pulses of equal duration using the graphical method.

EXAMPLE 3: PULSE TRANSMISSION OVER A TELECOMMUNICATIONS CHANNEL

We now apply the technique to a digital symbol being transmitted over a linear telecommunications channel. In many cases both the transmitted pulse and the impulse response of the channel are designed to be as nearly symmetric as possible. In such cases, as in

Example 2, the profile of the reversed impulse response is identical to the impulse response function itself, and we can simply imagine the channel impulse response and the pulse profile sliding over each other. At any instant in time, the system output (the received pulse) is given by the area under the curve generated by multiplying the two appropriately overlapping time functions. A technique often used in practice is for the transmitted pulse $g(t)$ and the overall channel impulse response $h(t)$ (by which are meant the channel itself together with any filters and equalisers) to have a very similar basic profile. The most important features of the received pulse $y(t)$ – its width, and the location of its maximum value – can be then deduced immediately by applying the graphical convolution method. Imagine sliding the two pulses over each other and multiplying. It should be clear from the preceding discussion that the width of the received pulse will be $2T$, and that its maximum value will occur at a time corresponding to the complete overlap of the convolved waveforms – that is, at time T. As would be expected, the received pulse has been 'smeared' over a time interval equal to the sum of the durations of the input pulse and the channel impulse response: this is illustrated in Fig. 2.44.

To quantify the precise shape of the output pulse, the convolution integral would need to be calculated, or the system modelled in some other way – in either case, probably by computer simulation. The qualitative result, however, is a classic illustration of how engineers use highly sophisticated mathematical models in a relatively simple and very practical way!

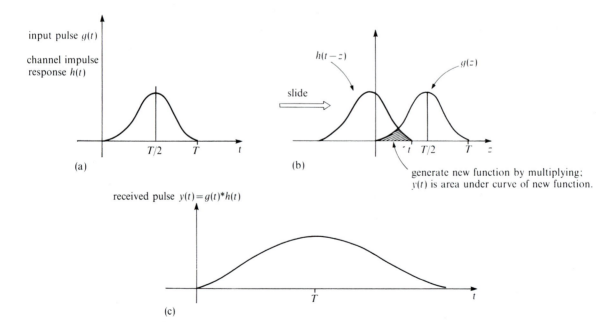

Fig. 2.44. The graphical method of convolution applied to the pulse response of a telecommunications channel.

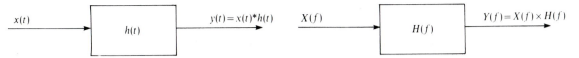

Fig. 2.45. Input–output behaviour of a linear system in time- and frequency-domains.

We now have two completely equivalent ways of specifying signals and the effects on them of linear systems. In the time-domain, the output of a linear system is given by the convolution of the input and the system impulse response. In the frequency-domain the spectrum of the output is given by multiplying the spectrum of the input and the system frequency response. This is summarised in Fig. 2.45: we could say that convolution in the time-domain is equivalent to multiplication in the frequency-domain. Which modelling approach we chose in any particular circumstances depends on various factors. What is important is to be able to move freely between the two domains, and most of the information needed to be able to do this for engineering purposes has been presented in this – admittedly rather long – chapter.

2.8 Summary

A number of techniques for modelling signals and systems have been presented. One of the most important types of model for telecommunications is the linear system, defined as a system which obeys the principle of superposition. A linear system can be fully specified by its frequency response function $H(f)$ or by its impulse response $h(t)$. A telecommunications channel whose amplitude ratio $|H(f)|$ is constant and whose phase characteristic $\theta(f) = \mathrm{Arg}\, H(f)$ is linear over the frequency range of interest is said to be distortionless.

Signals are often modelled as frequency spectra. A periodic signal has a line spectrum which specifies the amplitude and phase relationships of (a possibly infinite number of) discrete spectral components. A pulse $g(t)$ of finite duration has a spectrum consisting of a continuous distribution of components $G(f)$ defined by the Fourier transform

$$G(f) = \int_{-\infty}^{+\infty} g(t) \exp(-2\pi jft)\, \mathrm{d}t$$

where $g(t)$ and $G(f)$ are known as a Fourier transform pair and are often written simply as

$$g(t) \leftrightarrow G(f).$$

Various standard pulse models have been introduced, including: the rectangular pulse; the Dirac delta function or unit impulse; the triangular pulse; and the raised-cosine pulse. The rectangular pulse height V extending from $-\tau/2$ to $+\tau/2$ is a particularly important model, and has the spectrum

$$G(f) = V\tau \left(\frac{\sin \pi \tau f}{\pi \tau f} \right)$$

A number of general rules about time- and frequency-domains follow from an examination of the features of such standard models:

1 The shorter the pulse, the broader its spectrum.
2 Sudden changes in a pulse shape imply high frequencies in the corresponding spectrum.
3 Sudden changes in a spectrum imply oscillatory behaviour in the corresponding signal.

The impulse response $h(t)$ and the frequency response $H(f)$ of a linear system form a transform pair. The input $x(t)$, output $y(t)$ and impulse response $h(t)$ are related by the convolution operation

$$y(t) = x(t)*h(t) = \int_{-\infty}^{+\infty} x(z)\, h(t-z)\, dz$$

while the input and output spectra are related by the multiplication rule:

$$Y(f) = X(f) \times H(f).$$

Similar results hold for a linear discrete system processing an input sequence of samples $x(n)$ to produce an output sequence $y(n)$. A discrete linear system is fully characterised by its unit sample response sequence $h(n)$. In particular

$$y(n) = x(n)*h(n)$$

where the discrete convolution operation can be defined by:

$$y_i = \sum_{k=0}^{i} x_k h_{i-k}$$

At various stages in a digital telecommunication system analogue signals are sampled and processed in discrete form. A sampled signal can be modelled by a spectrum which consists of replicas of the continuous signal spectrum repeated about multiples of the sampling rate. If the frequency components present in a continuous signal extend from 0 to B Hz, then the signal can be completely represented by, and reconstructed from, a sequence of equally spaced samples, providing that the sampling frequency exceeds $2B$ samples per second. This important result is known as Shannon's sampling theorem.

3

Random signals and noise

3.1 Introduction

The signal models introduced so far (periodic signals and isolated pulses) are all *deterministic* – that is, their behaviour at any instant in time is completely specified in advance. Clearly, though, the signals encountered in telecommunications systems do not often behave like this. The precise sequence of digital symbols transmitted as a message is not known in advance – otherwise there would be no point in transmitting it! And as far as noise is concerned, although the *range* of voltages to be expected might be known, there is no way of predicting the *precise* noise level at any particular instant in time. Many waveforms in telecommunications therefore need to be treated as random variables, and modelled using statistical tools. This chapter introduces such tools.

3.2 Statistical averages (means)

In Chapter 2 the mean value of a periodic signal was identified with its zero-frequency or d.c. component. A similar idea applies to a random waveform. If the random signal is lowpass filtered so as to remove almost all the time-varying components, as illustrated in Fig. 3.1, then the result is a waveform which wanders only slightly from its mean value or d.c. level. The mean value of a random waveform is therefore equal to its d.c. or zero-frequency component.

A mathematical expression for the mean value follows immediately. Fig. 3.2 is an enlarged version of the input to the filter in Fig. 3.1. Over a given observation time the signal is equally likely to be greater or less than the mean, $\bar{x}$, so the total area under the curve must be equal to the

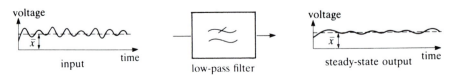

Fig. 3.1. Lowpass filtering reduces signal fluctuations about its mean value.

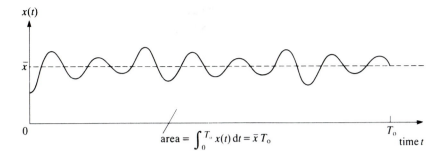

Fig. 3.2. The mean value of a signal expressed as a time integral.

rectangular area bounded by the time axis and the line representing the mean. For an observation time T_o we have

$$\bar{x} = \frac{1}{T_o} \int_0^{T_o} x(t)\, dt$$

Many random waveforms are observed to have mean values which do not change substantially over long periods of time. Then, the long-term average is given by

$$\bar{x} = \lim_{T_o \to \infty} \frac{1}{T_o} \int_0^{T_o} x(t)\, dt$$

The mean value is an important piece of information about a random variable, but of limited usefulness by itself since it gives no information about how far and how fast the signal fluctuates *about* the mean. One useful additional measure is the *mean-square value* of the signal. As its name implies, to find the mean-square value, the signal is first squared, and then the mean is taken of the result. Hence we have the long-term mean-square value:

$$\overline{x^2} = \lim_{T_o \to \infty} \frac{1}{T_o} \int_0^{T_o} x^2(t)\, dt$$

Note that the mean square is not the same as the square of the mean. This is illustrated for the simple case of a sinusoid in Fig. 3.3. The mean value of $\sin \omega t$ is zero, while the mean square is $1/2$.

The link between mean and mean square is provided by the *variance* of the signal. Variance may be defined as the mean-square deviation of the

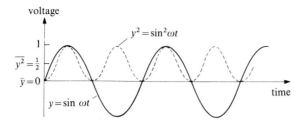

Fig. 3.3. The square of the mean is not, in general, equal to the mean of the square!

signal about its mean, and is given the symbol σ^2. Hence

$$\sigma^2 = \lim_{T_0 \to \infty} \frac{1}{T_0} \int_0^{T_0} [x(t) - \bar{x}]^2 \, dt$$

Like the mean value, mean square and variance also have useful physical interpretations. Suppose that a randomly-varying current $i(t)$ is applied to a $1\,\Omega$ resistor. The instantaneous power dissipated is i^2 and the mean power is therefore equal to the mean-square current. A similar argument applies to a mean-square voltage: either can be thought of as a measure of the mean power of a signal.

For a signal with zero mean, $\bar{x} = 0$ and the variance is identical to the mean-square value: in other words, the power of a signal with a.c. components only is numerically equal to the variance. This leads to a general interpretation of variance as the power in the a.c. components of a signal. Since the d.c. power is equal to the square of the mean, we can then write

$$\text{total power} = \text{d.c. power} + \text{a.c. power}$$
$$\text{mean square} = \text{square of mean} + \text{variance}$$
$$\overline{x^2} = (\bar{x})^2 + \sigma^2$$

In the vast majority of telecommunications applications noise is modelled as a random voltage or current with zero mean. In such cases the variance is equal to the mean square and is a direct measure of the noise power.

In addition to mean square or variance the square roots of these quantities, *root mean square (rms)* and *standard deviation (σ)*, respectively, are often used.

3.3 Probability density functions

The approaches just presented to mean, mean square and variance are based on a knowledge of the way the signal varies with time – in other words on a specific record or mathematical model of $x(t)$. However, it is often more convenient to work with what is known as the *probability distribution* or *probability density function* of the signal – that is, a representation of how likely it is that the signal will take on particular values. The probability distribution can be defined in the following way: If a random variable x has the probability distribution $p(x)$ then the probability of x taking a value between x_1 and x_2 is given by

$$\int_{x_1}^{x_2} p(x)\,dx$$

In other words, the probability of the variable taking on a value in a given range is equal to the corresponding area under the probability density curve, as shown in Fig. 3.4. Since the variable must take on *some* value, the total area under a probability density function must be 1.

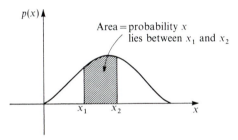

Fig. 3.4. A probability density function. The probability of x falling between x_1 and x_2 is equal to the shaded area.

Statistical averages can be calculated directly from a given probability density function. Consider first the mean value. The time integral presented earlier can be viewed as the limit of taking a large number of sample values at regular time intervals, adding them together, and dividing by the total number of samples in the usual way. This process is illustrated in Fig. 3.5(a). The alternative procedure using the probability density function is illustrated in part (b) of the figure. Suppose we take a large number N of samples. The probability of the signal taking on a value within a small range δx about x is $p(x)\delta x$, so the actual number of samples lying within this range is $Np(x)\delta x$. To calculate the mean we have to sum the total values of all the samples in all the ranges and divide by N. In the range δx, the $Np(x)\delta x$ samples each have a value very close to x, so the sum of values in this range is very close to $Nxp(x)\delta x$. Letting N become indefinitely large and δx indefinitely small, and summing over all possible ranges leads to the expression:

$$\text{mean} = \bar{x} = \lim_{\substack{N \to \infty \\ \delta x \to 0}} \frac{\sum\limits_{-\infty}^{\infty} N x p(x) \delta x}{N}$$

$$= \int_{-\infty}^{+\infty} x p(x) \, dx$$

The probability
distribution approach
may seem rather
complicated, but it has
the important
advantage of providing
a direct link between
the overall statistics of
a random signal and
the important average
measures of the signal.

In a precisely similar way we have

$$\text{mean square} = \overline{x^2} = \int_{-\infty}^{+\infty} x^2 p(x) \, dx$$

and

$$\text{variance} = \sigma^2 = \int_{-\infty}^{+\infty} (x - \bar{x})^2 p(x) \, dx$$

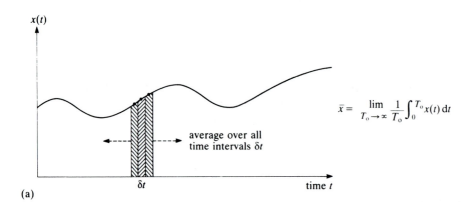

$$\bar{x} = \lim_{T_0 \to \infty} \frac{1}{T_0} \int_0^{T_0} x(t) \, dt$$

(a)

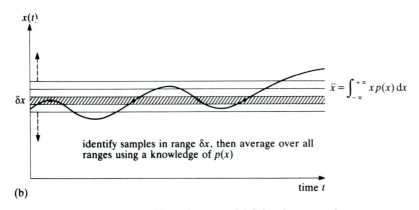

$$\bar{x} = \int_{-\infty}^{+\infty} x p(x) \, dx$$

(b)

Fig. 3.5. Alternative ways of defining the mean value.

EXAMPLE

A random signal has the probability density function shown in Fig. 3.6 – that is, any voltage between 1 and 5 mV is equally likely, and the signal never takes on a value outside that range. What are the mean, mean square, variance and standard deviation of the signal, and what total power would be dissipated by the signal in a 1 Ω resistor?

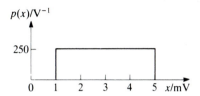

Fig. 3.6. A rectangular probability density function. The variable x is equally likely to take any value between 1 and 5 mV, and never falls outside this range.

Solution. By symmetry, the mean value is 3 mV, since the random signal is equally likely to take on a value above or below the mean. Alternatively,

$$\bar{x} = \int_{-\infty}^{+\infty} x p(x)\, dx$$

The limits here are simply the range of possible values of x, while $p(x)$ is a constant $250\,\mathrm{V}^{-1}$ over this range:

$$\bar{x} = 250 \int_{10^{-3}}^{5 \times 10^{-3}} x\, dx$$

$$= 250 \left[\frac{x^2}{2} \right]_{10^{-3}}^{5 \times 10^{-3}}$$

$$= 3 \times 10^{-3}\,\mathrm{V}$$

$$= 3\,\mathrm{mV}$$

$$\text{mean square} = \overline{x^2} = \int_{-\infty}^{+\infty} x^2 p(x)\, dx$$

$$= 250 \left[\frac{x^3}{3} \right]_{10^{-3}}^{5 \times 10^{-3}}$$

$$= 10.33 \times 10^{-6}\,\mathrm{V}^2$$

The variance is most easily calculated by using the expression

$$\text{mean square} = \text{variance} + \text{square of mean}$$

Hence for the variance we have

$$\begin{aligned}
\sigma^2 &= \overline{x^2} - (\bar{x})^2 \\
&= (10.33 - 9) \times 10^{-6} \\
&= 1.33 \times 10^{-6} \, \text{V}^2
\end{aligned}$$

Standard deviation is equal to the square root of the variance, giving

$$\sigma = 1.15 \times 10^{-3} \, \text{V} = 1.15 \, \text{mV}$$

Finally, the total power dissipated in a $1\,\Omega$ resistor is numerically equal to the mean-square voltage already calculated. Hence the total power $= 10.33 \, \mu\text{W}$.

EXERCISE 3.1

A source of random noise can take on a value between 0 and V volts only, and all values within this range are equally likely. Derive an expression for the total noise power, assuming dissipation in a $1\,\Omega$ resistor.

3.3.1   The Gaussian distribution

The type of probability distribution of the previous example, in which all values within a tightly defined range are equally probable, is useful for modelling certain random signals encountered in telecommunications – the error introduced by digitisation, for example, as will be discussed in Chapter 7. The most common types of noise, however, have a probability distribution very different from this: certain ranges are much more likely than others, and there is not usually such a clear limit to the possible noise voltages which may occur.

A model which is found to approximate quite closely many physical noise processes is the *Gaussian* or *normal distribution*, illustrated in Fig.

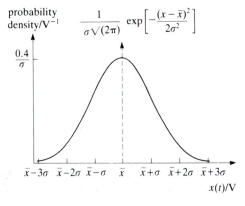

Fig. 3.7. The Gaussian distribution.

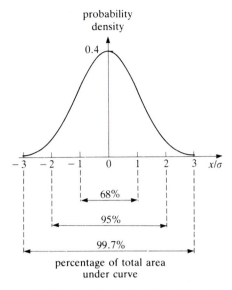

Fig. 3.8. The 'normalised' Gaussian distribution. Note that each axis is labelled in dimensionless units.

This is known in probability theory as the Central Limit Theorem.

3.7. One reason for this is that a Gaussian distribution is approached when a large number of random processes with identical distributions combine to produce a cumulative effect – even if the individual processes are not themselves Gaussian. So, for example, the random motion of many charge carriers in an electronic circuit often gives rise to Gaussian noise.

Fig. 3.7 shows how a Gaussian distribution is characterised by its mean and standard deviation σ. It is often more convenient to work with a slightly different plot, as shown in Fig. 3.8, where the random variable is assumed to have zero mean, and each axis is dimensionless. Note in particular that the horizontal axis is labelled in multiples of the standard deviation. As with all probability distributions, the area under a Gaussian distribution between two values of the random variable represents the probability of the variable lying within the corresponding range: the total area under the curve is unity. Fig. 3.8 also includes some useful rules of thumb: the probability of the random variable lying within $\pm\sigma$ of the mean is 0.68; within $\pm2\sigma$, 0.95; and within $\pm3\sigma$, 0.997.

EXAMPLE

A Gaussian random voltage has a mean value of 6 V and a standard deviation of 1.5 V. What is the probability that the signal will lie in the range 4.5–9.0 V?

Solution. The range is from σ below the mean to 2σ above the mean. There is a probability of $\frac{0.68}{2}=0.34$ that the signal voltage is between the mean and σ *below*, and $\frac{0.95}{2}=0.475$ that it will lie between the area and 2σ *above*. Hence the total probability is $0.34+0.475=0.815$.

3.4 Frequency-domain characterisation of random signals

Fig. 3.9 shows a segment of a random waveform. Its amplitude and phase spectrum could be calculated by means of the Fourier transform to give an equivalent, and perfectly valid, frequency-domain model. However, if the procedure were then repeated with another segment of the waveform the spectrum could be very different, and in general it would be unpredictable. To characterise a random signal in the frequency domain a different measure is used, one which depends only on the *average* frequency-domain properties of the signal.

Suppose that a randomly-varying signal is passed through a narrow, tunable, bandpass filter and the power of the output is measured. (Power can be measured in a number of ways – for instance, by sampling the filter output and calculating its mean-square value.) The power in the individual frequency bands can then be plotted against frequency, as shown in the histogram or bar chart of Fig. 3.10(a). Suppose now that the bandwidth of the filter is halved, keeping its other characteristics constant, and the procedure repeated. The result might be as in Fig. 3.10(b): the same general shape, but the narrower frequency increment

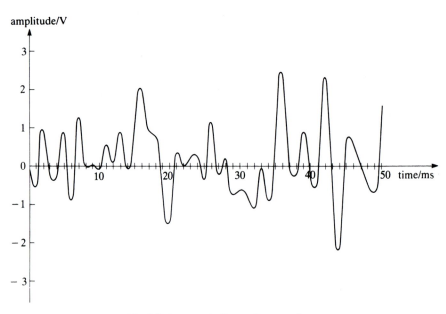

Fig. 3.9. A segment of a random waveform.

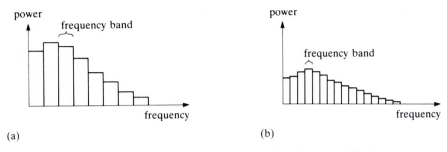

Fig. 3.10. Power in a given frequency band as a function of frequency. The frequency band in (a) is twice the width of that in (b).

This procedure for characterising the spectrum of a random signal will be meaningful only if the signal is what is known as *stationary* – that is, its statistical properties are effectively unchanging with time. Most random waveforms commonly encountered in telecommunications (e.g. thermal noise) can be assumed to be stationary.

has improved the fine detail. To obtain an accurate measure of the distribution of power over the frequency range, the frequency band needs to be reduced indefinitely. To prevent the heights of the individual bars becoming indefinitely small, it is usual to plot the power density (in watts per hertz) against frequency, instead of the actual power in a given finite frequency band. Then, as with the other density functions introduced so far, it is the *area* under the curve which is significant, as shown in Fig. 3.11. The area between two frequencies under a *power density spectrum* (also called *power spectral density*, and often abbreviated to *power spectrum*) represents the power in that frequency range.

Like the amplitude and phase spectra introduced in Chapter 2, power densities can be plotted in single- or double-side form. In both forms the area under the curve is equal to the power, so numerical values of power density in single-sided form are twice the corresponding ones in the double-sided representation. Power spectral densities are usually given the symbols S or P and are commonly expressed as functions of either frequency or angular frequency. It is also common to use mean-square voltage or current as a direct measure of power, so power spectra are often quoted as 'volts-squared per hertz' or 'amperes-squared per hertz'. Finally, note that the square roots of these measures are also sometimes used, leading to the rather bizarre-sounding units of rms volts (or amps) per *root* hertz.

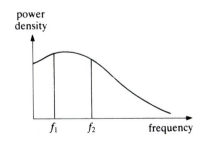

Fig. 3.11. A continuous power density spectrum.

Although power spectra are particularly useful for characterising random signals and noise, any practical signal with finite power has a power spectrum, regardless of whether it is random. (In signal processing it is common to distinguish between signals with *finite energy*, such as individual pulses, and those with *finite power*, such as an indefinitely long pulse train, or a source of noise.) A periodic signal has a line power spectrum, in which the heights of the individual lines represent the powers (not the amplitudes) of the individual components; a finite-power signal without any periodic components has a continuous power spectrum. In all cases, though, a power spectrum measures *average* frequency-domain properties. It contains no phase information, and it is impossible to deduce the shape of a signal as a function of time from a knowledge of its power spectrum alone.

EXAMPLE

A telephone channel suffers from Gaussian noise which can be modelled by a constant single-sided power density of $10^{-6} \, V^2 \, Hz^{-1}$ over the frequency range 400 Hz–4 kHz. Bandpass filters in the system ensure that the noise outside this range is negligible. What is the total noise power?

Solution. The total power of the noise is given by the area under the power spectrum, that is

$$\begin{aligned} \text{noise power} &= 10^{-6} \times 3600 \, V^2 \\ &= 3.6 \times 10^{-3} \, V^2 \end{aligned}$$

If dissipated in a $1 \, \Omega$ resistor, the power is therefore 3.6 mW.

EXERCISE 3.2

What is (a) the mean and (b) the rms value of the noise in the previous example.

3.4.1   White noise

A useful spectral model of many types of noise encountered in practice is *white noise*. White noise has a constant power spectral density for all frequencies: its power spectrum is just a horizontal line, as shown in Fig. 3.12(a). Theoretically the spectrum extends from d.c. to infinity, so the total noise power without filtering appears to be infinite. Any naturally occurring noise, however, must roll-off at sufficiently high frequencies, so

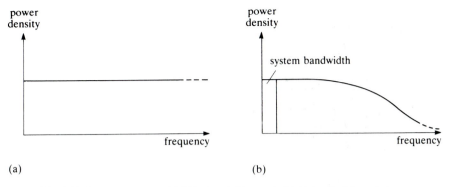

Fig. 3.12. Power spectrum of (a) ideal and (b) practical wideband white noise.

a more realistic model, with finite total noise power, is as shown in Fig. 3.12(b). All telecommunication systems have a finite bandwidth, and system frequency response often cuts off far below the point at which the noise begins to roll-off. The white noise model, characterised by a constant spectral density *for all frequencies of interest*, is therefore a perfectly valid modelling tool.

In the absence of an applied emf, the electrons in a resistor are in purely random motion, their agitation increasing with temperature. This random motion leads to a randomly-varying potential difference across the resistor, known as *thermal* or *Johnson noise*.

A resistor generating thermal noise can be represented by a noise voltage generator in series with an ideal, noiseless resistor of the same resistance, as shown in Fig. 313(a). Then the noise can be closely modelled as white, Gaussian noise with a single-sided power spectral density of $4kTR \, \text{V}^2 \, \text{Hz}^{-1}$, where T is the absolute temperature in kelvins and k is Boltzmann's constant ($k = 1.38 \times 10^{-23} \, \text{JK}^{-1}$).

A completely equivalent circuit is shown in Fig. 3.13(b), where the noisy resistor is represented by an ideal, noiseless one in parallel with a noise current generator. Then the single-sided power spectral density is $4kT/R \, \text{A}^2 \, \text{Hz}^{-1}$. A quantum mechanical analysis shows that thermal noise roll-off begins at around 10^{15} Hz or higher, depending on temperature. This is well beyond the response of any conventional electronic component, so the white noise model is usually perfectly adequate over the bandwidth of most telecommunication systems.

Another physical source of noise which can often be treated as white and Gaussian is *shot noise*. Shot noise arises when a discrete charge carrier crosses a potential boundary in an electronic device such as a diode, transistor, or vacuum tube. When charge carriers cross a boundary such as a semiconductor junction, random variations in the precise rate of flow of carriers across the boundary give rise to random fluctuations in the

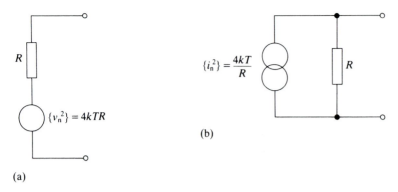

Fig. 3.13. (a) Thévenin and (b) Norton equivalent circuits for a noisy resistor.

current flow. The effect can be modelled as a shot noise current with a single-sided power spectral density of $2qI$ amps-squared per hertz, where q is the carrier charge and I is the ideal current upon which the shot noise current is superimposed. If, as is usually the case, the carriers are electrons, then $q = 1.6 \times 10^{-19}$ C. Provided the current I is sufficiently large, then shot noise can be treated as Gaussian; this will often, but not always, be the case.

3.4.2   Coloured noise

Although many physical noise sources can be accurately modelled as white, after the noise is filtered (by an equaliser in a digital receiver, for example) it no longer has a constant spectral density, but takes on the spectral characteristics of the filter. Such noise is termed *coloured*.

This is illustrated in general terms in Fig. 3.14, which shows the low-pass filtering of noise and also introduces a new notation $\{v^2\}$ for mean-square voltage spectral density. In general $\{v^2\}$ will be a function of frequency, but for white noise it will be constant, and the symbol η will be used instead. (The $\{v^2\}$ notation is not standard, but it will be found to simplify some of the calculations in Chapter 9.)

The relationship between input and output mean-square voltage densities for a linear system is very similar to the results presented earlier for individual sinusoids and continuous spectra. For input and output sinusoids V_i and V_o respectively, we had

$$V_o = V_i \times |H(f)|$$

while for continuous spectra

$$|G_o(f)| = |G_i(f)| \times |H(f)|.$$

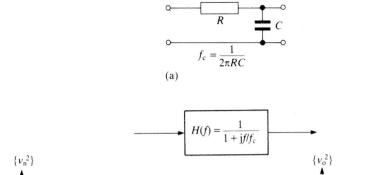

Fig. 3.14. Passing wideband white noise through a lowpass filter.

Since we are now dealing with *mean-square* quantities, we have

$$\{v_o{}^2\} = \{v_i{}^2\} \times |H(f)|^2$$

or, using the alternative notation for general power spectra

$$S_o(f) = S_i(f) \times |H(f)|^2$$

Applying this result to the filtering of white noise with constant mean-square voltage density η gives an output noise power spectrum

$$\{v_o{}^2\} = \eta |H(f)|^2$$

where $H(f)$ is the frequency response function of the filter. To calculate the resulting noise *power* (mean-square noise voltage) at the filter output we must compute the area under the output power spectrum. The following example illustrates this for first-order low-pass filtering with

$$H(f) = \frac{1}{1 + jf/f_c}$$

EXAMPLE

Wideband white noise with a single-sided spectral density of $2 \times 10^{-8} \, V^2 \, Hz^{-1}$ is filtered by a first-order low-pass filter with unity d.c. gain and a 3 dB cut-off frequency of 1 kHz. What is the mean-square noise voltage at the filter output?

Solution. The output power spectrum is given by

$$\{v_o{}^2\} = \frac{\eta}{1 + (f/f_c)^2}$$

The mean-square value is obtained by calculating the area under the curve for all frequencies – that is, by integrating over $0 < f < \infty$.

$$\overline{v_o^2} = \eta \int\limits_0^\infty \frac{f_c^2}{f_c^2 + f^2} \, \mathrm{d}f$$

From standard integral tables (or by making an appropriate trigonometrical substitution) we have

$$\overline{v_o^2} = \eta f_c \left[\tan^{-1} \frac{f}{f_c} \right]_0^\infty = \eta \frac{\pi}{2} f_c$$

Substituting in values for $\eta = 2 \times 10^{-8} \, \mathrm{V}^2 \, \mathrm{Hz}^{-1}$ and $f_c = 10^3 \, \mathrm{Hz}$ gives

$$\overline{v_o^2} = 3.14 \times 10^{-5} \, \mathrm{V}^2$$

EXERCISE 3.3

Suppose that instead of using a first-order low-pass filter, the noise is passed through a filter closely approximating an ideal 'brick-wall' low pass filter with cut-off 1 kHz. Compare the output mean-square voltage with the previous example.

3.4.3   Power spectra of random digital signals

As was noted at the beginning of this chapter, information-bearing signals exhibit many of the characteristics of randomness. It is therefore often useful to have a knowledge of the general shape of the power spectrum of the pulse train corresponding to a random binary message sequence of zeros and ones. In general, the spectrum of such a pulse train will depend both on the type of coding used and the shape of the individual pulses transmitted. The effects of various coding schemes will be considered in a later chapter. Here we concentrate on the consequences of the randomness of the bit pattern, modelling the generation of pulses as described in Section 2.5.2. That is, we imagine the pulses to be generated by applying unit impulses to a filter whose frequency response $G(f)$ is identical to the spectrum of the desired pulse $g(t)$. A random sequence of *pulses* is therefore generated by applying an appropriate random sequence of *impulses* to the pulse generating filter. The power spectrum of the filter output can be derived from a knowledge of the power spectrum of the input using the relationship

$$S_o(f) = S_i(f) \times |G(f)|^2$$

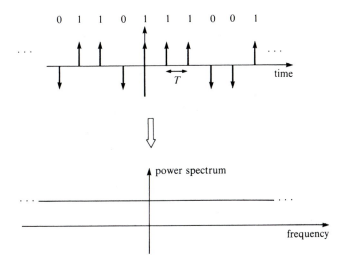

Fig. 3.15. A bipolar, random, binary impulse train and its power spectrum.

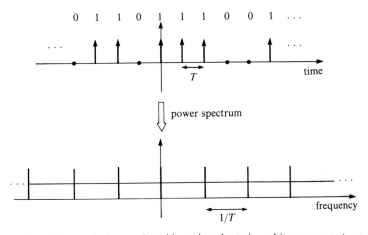

Fig. 3.16. A unipolar, random, binary impulse train and its power spectrum.

providing we know the power spectrum of the random, input impulse train.

The mathematical derivation of the spectra of impulse trains is beyond the scope of this book, but two important results are presented, without proof, in Figs. 3.15 and 3.16. The first is the spectrum of a random, *bipolar* impulse train, where unit positive and negative impulses are equally likely. The power spectrum is a constant value for all frequencies: given the flat amplitude spectrum of an isolated impulse (Section 2.4.2) this is perhaps what might have been expected. Fig. 3.16 shows the corresponding power spectrum for a *unipolar* impulse train (one binary symbol represented by a unit impulse, the other by the absence of an impulse).

Note that there is again a constant, continuous element in the power spectrum but that in addition we have discrete spectral lines at zero frequency (d.c.) *and all multiples of the sampling frequency* $1/T$. Since the impulses are all of the same polarity, the presence of a d.c. term is only to be expected, but the spectral lines may come as more of a surprise. One way of viewing this latter feature is that they are a consequence of what might be termed the 'periodic' aspect of such impulse trains. Impulses can occur only at periodically repeating time intervals (integral multiples of T), and this fact is reflected in the presence of discrete spectral components at frequencies which are integral multiples of $1/T$. It turns out that in the bipolar case, the presence of, on average, an equal number of positive and negative impulses means that all such discrete components in the spectra (both d.c. and higher-frequency) are eliminated.

Another way of looking at such power spectra is to draw an analogy with sampling. The spectra of such impulse trains and those of sampled signals repeat at intervals equal to the sampling frequency. Readers interested in following up the detailed mathematical background are referred to Schwartz or Meade & Dillon.

EXAMPLE

Sketch the general form of the power spectrum of a random, bipolar, binary signal using NRZ rectangular pulses of width T. How does this spectrum differ from that of a similar, unipolar signal?

Solution. The input–output relationship for power spectra implies that the power spectrum will have the general form

$$\text{constant} \times |G(f)|^2.$$

For a square pulse, height V, width T, $G(f) = VT (\sin \pi Tf)/\pi Tf$, so the general form of the power spectrum of the random bipolar signal is as shown in Fig. 3.17(a).

The power spectrum of a random, unipolar NRZ signal will have a similar continuous component with a general $|(\sin x)/x|$ squared shape, but in addition there will be a discrete line at d.c. (correspond-

In the absence of a detailed mathematical treatment quantifying the height of the power spectrum of the input impulse train, the vertical axis of Fig. 3.17 cannot be labelled with numerical values. Much more important from the engineering point of view, however, are other features of the spectrum such as the relative heights of the side lobes and the nulls in the spectrum at frequencies which are multiples of $1/T$.

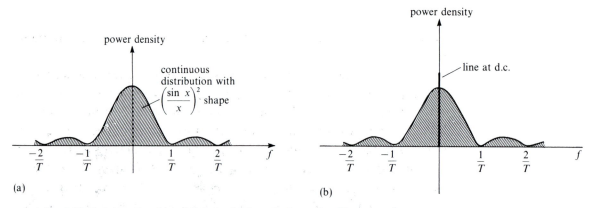

Fig. 3.17. Power spectra of (a) a bipolar and (b) a unipolar random binary waveform using full width (NRZ) rectangular pulses.

ing to the d.c. power of the waveform) as shown in Fig. 3.17(b). Note that no other spectral lines are present, since the impulse train spectral lines correspond to pulse spectral nulls.

EXERCISE 3.4

Sketch the general forms of the power spectra of a random binary message signal (with equiprobable zeros and ones) coded using

(a) bipolar; and
(b) unipolar

RZ (return-to-zero) *half-width* rectangular pulses.

Assume a signalling rate of $1/T$ states s^{-1}, and a pulse width of $T/2$. Distinguish clearly between continuous and discrete components in the spectra.

3.5 Combining random sources

In a telecommunications system there will normally be a number of different sources of noise. In many such cases it is the combined power (mean-square value) of all the sources which is of particular interest. Consider first two random sources, $x(t)$ and $y(t)$ which add together in a system. The total mean-square value is

$$\overline{(x+y)^2} = \overline{x^2} + \overline{y^2} + \overline{2xy}$$

Writing the average product $\overline{xy}$ as a time integral gives

$$\overline{xy} = \lim_{T_0 \to \infty} \frac{1}{T_0} \int_0^{T_0} x(t)y(t)\,dt$$

This expression can be interpreted as a measure of the 'similarity' or 'correlation' between $x(t)$ and $y(t)$. If x and y are identical, then the integral reduces to the mean-square value of either. If, on the other hand, the two sources are sufficiently 'different', then the long-term average of the product of x and y approaches zero. Two such random sources with this property – or indeed any two such finite power signals – are said to be *orthogonal*. For orthogonal waveforms

$$\overline{xy} = 0$$

and

$$\overline{(x+y)^2} = \overline{x^2} + \overline{y^2}$$

In other words, we can add the mean-square values (powers) of individual orthogonal, finite-power, sources to obtain the overall mean-square value (power). The detailed mathematical background to the subject of orthogonality need not concern us here: it is sufficient to note that *independent random sources with zero mean values are orthogonal*. Separate physical sources of noise, such as individual resistors, transistors and so on, can usually be modelled as independent with zero mean, so the total noise power can usually be obtained as an appropriate sum of individual contributions. Similarly, the overall noise power spectrum can be obtained by simply adding together the spectra of the individual sources.

Note that separate noise waveforms cannot be assumed to be orthogonal unless we are certain that they arise from statistically independent physical processes. Fig. 3.18 shows an extreme example where this is not the case: the two waveforms originate from the same noise source and combine at a later stage. Then the two signals are *not* independent, and individual powers *cannot* be added to obtain the combined power. In the example shown, a random signal is inverted and added to a version of itself. Far from adding powers, the two signals cancel out completely: such techniques can actually be used for acoustic noise cancellation.

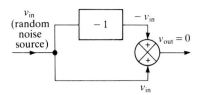

Fig. 3.18. An example where two random signals cancel.

The concept of orthogonality can also be applied to finite energy signals, such as the individual pulses used in digital signalling. This topic will be taken up in Chapter 4, where a new and powerful model of digital signals will be developed, based on such ideas.

3.6   Noise and threshold detection in digital systems

At some point in a digital receiver regular decisions are made about the state of the incoming waveform – that is, decisions about which of the permitted symbols the incoming signal represents. In the simplest case, this could be direct threshold detection of the received voltage level; more commonly it would be threshold detection after some sort of equalisation and/or other signal processing to emphasise the distinction between the expected symbols. In either case, however, noise can shift the level of the detected signal at the sampling instant to the wrong side of the threshold, causing a misinterpretation of the received signal. The likelihood of such an error occurring will depend on the probability density function of the noise.

Consider first a baseband, unipolar, binary system using zero and V_{trans} volts as the two transmitted states, such that a detected voltage level at the receiver greater than a given threshold V_t is interpreted as a binary 1, and a voltage less than V_t as a binary 0. As mentioned above, the noise encountered in practice can often be closely approximated as Gaussian. Even when it cannot, mathematical results assuming Gaussian noise are useful idealisations leading to important theoretical limits to perform-ance. A Gaussian model will therefore be assumed here.

3.6.1   The Q-function

For the system under consideration it is the probability of the noise voltage exceeding a given value which is of significance. Suppose, for example, that zero volts is transmitted, corresponding to a binary 0. The voltage detected will be entirely due to noise, and the probability of the symbol being misinterpreted as binary 1 will be equal to the probability of the noise voltage exceeding the threshold. This is illustrated in Fig. 3.19 for noise of zero mean, as is usually the case. In part (a) of the figure, the noise has low rms value (equal to the standard deviation for zero mean), and there is negligible probability that the threshold will be exceeded. In part (b) however, the tail of the Gaussian curve extends well beyond the threshold: the probability of error is equal to the shaded area.

Fig. 3.20 is the analogous situation for a binary 1, when a detected V volts would be expected (less than V_{trans} owing to attenuation in the channel). In this case an error will occur if the noise voltage is sufficiently *negative* to bring the total *below* the threshold. Again, the error probability is equal to the shaded area under the tail.

Because of the importance of the area of this tail of a Gaussian distribution, its value is tabulated or plotted in mathematical tables as the so-called *Q-function*. Fig. 3.21 shows a Gaussian distribution in the standard form introduced earlier, and with a highlighted tail. The value of $Q(z)$ for any value of z is defined as the area of this tail. For example, $Q(2)$ is the probability that $x/\sigma > 2$ – that is, the probability that the random variable will have a (positive) value more than twice its standard

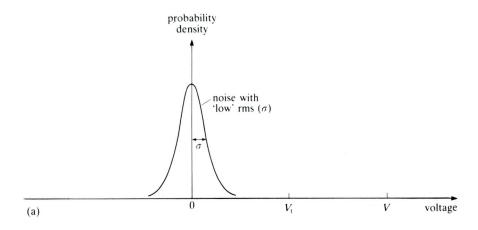

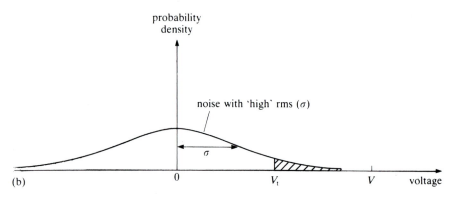

Fig. 3.19. Probability density function of received voltage when binary 0 is transmitted: (a) noise with low rms, (b) noise with high rms.

deviation. To obtain the probability that a noise value with zero mean exceeds a given threshold V_t, the value of $Q(V_t/\sigma)$ must therefore be found. Some values of $Q(z)$ are tabulated in Table 3.1, while a plot of the function is reproduced as Fig. 3.22.

Table 3.1. *Values of* $Q(z)$

z	$Q(z)$
1	0.16
2	0.023
3	0.001 3
4	0.000 032
5	0.000 000 29

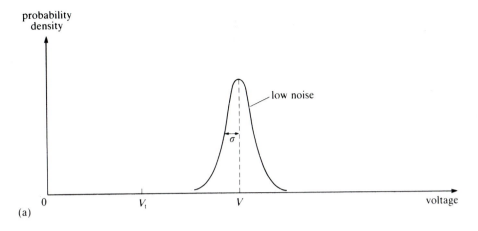

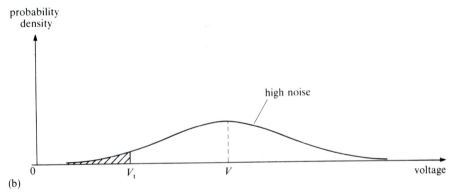

Fig. 3.20. Probability density function of received voltage when binary 1 is transmitted:
(a) noise with low rms. (b) noise with high rms.

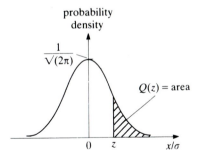

Fig. 3.21. Definition of the Q-function. $Q(z)$ is equal to the shaded area.

A useful close approximation to $Q(z)$ for values of z greater than 4 is given by the expression

$$Q(z) = \frac{\exp\left(\frac{-z^2}{2}\right)}{z\sqrt{(2\pi)}}$$

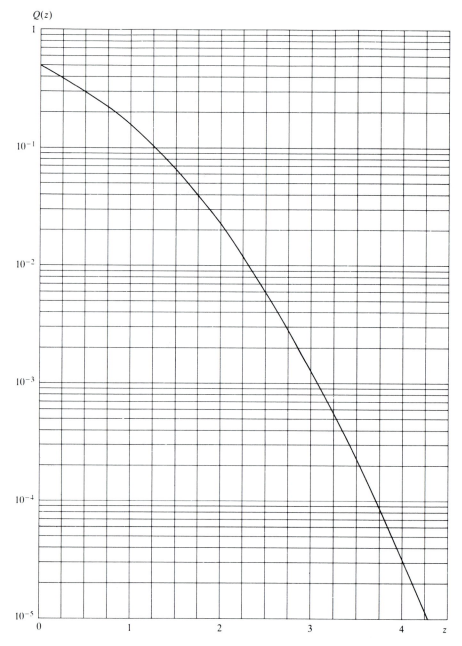

Fig. 3.22. The graph of $Q(z)$.

The Q-function can be applied just as easily to a random variable with non-zero mean. Then $Q(z)$ is the probability that the variable will have a value more than z times σ above the mean. And finally, because of symmetry, the Q-function can also be used for regions below the mean.

EXAMPLE

(a) A source of noise is Gaussian, with zero mean and rms value
0.2 V.
  (i) Use Table 3.1 to find the probability that the noise voltage
  will exceed 1 V.
  (ii) Use Fig. 3.22 to estimate the probability that the noise
  voltage will exceed 0.35 V.
(b) A second noise source is Gaussian with a mean of 0.4 V and a
standard deviation of 0.15 V. For what percentage of the time
would you expect the measured noise voltage to exceed 0.7 V?

Solution:

(a) 1 V is 5σ above the mean, so the probability of exceeding this
voltage is $Q(5)=0.000\,000\,29$.

   0.35 V is 1.75σ above the mean. Reading from Fig. 3.22,
   $Q(1.75)\approx 4\times 10^{-2}=0.04$.
(b) 0.7 V is 0.3 V (2σ) above the mean of 0.4 V. The probability of the
noise exceeding his voltage is therefore $Q(2)=0.023$, so this level
would be exceeded for 2.3% of the time.

EXERCISE 3.5

A telephone channel suffers from white Gaussian noise with a power
density of $1.5\times 10^{-6}\,\text{V}^2\,\text{Hz}^{-1}$ over the frequency range 400–3400 Hz.
There is negligible noise outside this range. What is the probability
that the noise voltage will exceed 100 mV?

3.6.2   Error rates

To estimate the overall probability of an error occurring it is necessary to
know the probability of binary 1s and 0s in the message. Then the overall
error probability P_e is given by

$P_e=$ (probability of binary 0 × probability of misinterpreting a 0)
$\quad+$ (probability of binary 1 × probability of misinterpreting a 1)

If binary 1s and 0s are equally likely – that is, the probability of either is
0.5 – then, in the presence of additive Gaussian noise, the overall error
probability is minimised by placing the threshold halfway between the
expected levels representing the two states so that in the above examples
$V_t = V/2$. This was tacitly assumed in Fig. 3.19 and 3.20.

Suppose as before that binary 0 is represented by zero volts and binary 1 by a received V volts. Then, for an error in a binary 0 we have

$$P_{e0} = \text{probability (noise voltage} > V_t)$$

that is, the area of the tail in Fig. 3.19.

Similarly, for a binary 1

$$P_{e1} = \text{probability (noise voltage} < -V_t)$$

that is, the area of the tail in Fig. 3.20.

Both of these areas are equal to $Q(V_t/\sigma) = Q(V/2\sigma)$ and the overall probability of error is hence

$$P_e = 0.5 \times Q(V/2\sigma) + 0.5 \times Q(V/2\sigma)$$
$$= Q(V/2\sigma)$$

This expression can be rewritten in terms of the signal to noise ratio of the system. There are various common definitions of signal to noise ratio. The most appropriate version here relates the peak signal level to the noise level, and can be written as either a 'voltage' or 'power' ratio. The *peak signal to noise voltage ratio* is defined as

$$(S/N)_v = \text{peak signal voltage/rms noise voltage}$$

while the 'power' ratio is

$$(S/N)_p = (\text{peak signal voltage})^2/\text{mean-square noise voltage}$$

$$\text{Hence } (S/N)_p = (S/N)_v^2$$

For the simple binary system under consideration the peak signal voltage is V, and the rms noise voltage is σ. Hence $(S/N)_v = V/\sigma$ and the expression $P_e = Q(V/2\sigma)$ can be rewritten as

$$P_e = Q\left[\frac{(S/N)_v}{2}\right]$$

Error probabilities for a unipolar binary system can hence be plotted directly in terms of signal to noise ratio, as shown in Fig. 3.23. Note the existence of a threshold region at around 10–12 dB. Beyond this point small improvements in S/N give spectacular improvements in error rate.

Fig. 3.23 can also be used for a bipolar binary system where $\pm V/2$ are the two received states instead of V and 0. In such a system, an error will still occur if the noise voltage exceeds $+V/2$ for one state, or is less than $-V/2$ for the other. The overall error probability is still therefore $Q(V/2\sigma)$, and Fig. 3.20 still applies, providing that the horizontal axis is still interpreted as $20\log(V/\sigma)$. (The error rates are the same because the

error rate

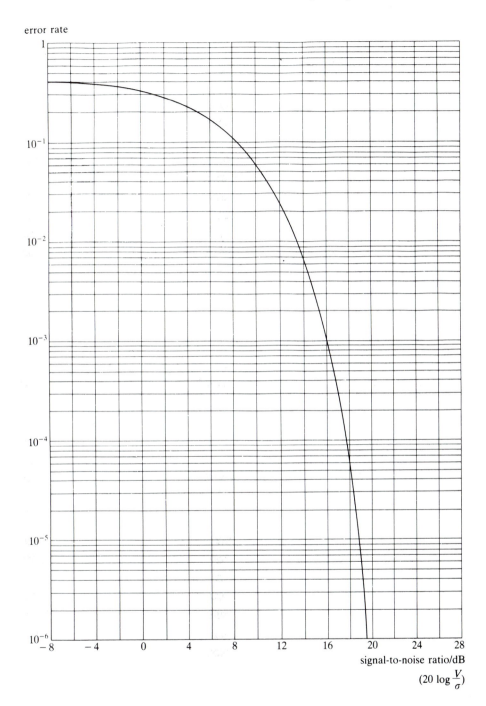

signal-to-noise ratio/dB

$(20 \log \frac{V}{\sigma})$

Fig. 3.23. Error rate as a function of S/N for unipolar binary signalling.

two states are 'the same distance (V volts) apart' a notion which will be clarified below.) Note, however, that $(S/N)_v$ is $V/2\sigma$ for the bipolar scheme, as opposed to V/σ for unipolar signalling, so the numerical values on the horizontal axis do not represent bipolar S/N in decibels.

Bipolar signalling usually requires only half the mean transmitted power of unipolar signalling for the same error rate: twice as many bipolar pulses will be transmitted, but each has only half the voltage of the unipolar pulses. Assuming that binary ones and zeros are equally likely, then the average transmitted power for unipolar full-width signalling is proportional to $V_{trans}^2/2$, while for bipolar it is proportional to $(V_{trans}/2)^2 = V_{trans}^2/4$.

EXERCISE 3.6

A digital system uses $+5$ V and -5 V as the two digital states, and a threshold of 0 V. Assuming that the noise at the receiver is additive and Gaussian, what is the maximum permissible rms noise voltage if overall error rate is to be kept below 1 in 1000?

3.6.3   Quantifying errors

The performance of a digital communication system is quantified primarily in terms of the expected statistics of the errors introduced by the link. The simplest and most widely used measure is the mean probability of any one bit being in error (P_e). This is just the ratio between the mean number of errors in any given interval and the total number of bits in that interval. It is often referred to as the *Bit Error Ratio* (*BER*).

The BER however, gives no information about the error distribution. The implicit assumption is often that the errors are being caused by Gaussian noise, as discussed in the previous section. This would lead to a binomial type distribution, in which the probability of a block of n bits having r errors is given by

For further information on the binomial distribution see any standard text on probability or the appendix in Schwartz.

$$P(r) = {}^nC_r p^r q^{n-r}$$

where p is the probability of any one bit being in error, $q = (1-p)$ is the probability of any one bit *not* being in error and ${}^nC_r$ is the binomial coefficient

$$^nC_r = \frac{n!}{r!(n-r)!}$$

In practice, however, the main cause of errors on a digital systems is often not Gaussian noise, but interference or impulsive noise which can give rise to errors in a comparatively large number of consecutive bits. Such error 'bursts' can make a considerable difference to the performance of the link. In pulse code modulation (described in Chapter 7), for example, Gaussian noise might give rise to a continuous slight background 'crackle', whereas the same BER appearing as bursts would completely obscure the speech for short periods of time. The distribution of errors also affects the success of error-correcting channel codes (see Chapter 6) in correcting introduced errors. In fact, some codes are designed to protect against errors with binomial-type distributions, others protect specifically against bursts.

In order to quantify the error distribution, CCITT have recommended the following parameters (CCITT blue books, recommendation G.821):

In practice, some users have not found degraded minutes to be a very useful measure, and the parameter may be dropped from the next edition of the CCITT recommendations.

degraded minutes – available minutes during which the average error ratio (after excluding severely errored seconds) is worse than 10^{-6}

severely errored seconds – available seconds during which the error ratio is worse than 10^{-3}

errored seconds – available seconds during which there are any errors (seconds which are not *error-free seconds*)

All these parameters are measured in what is known as *available time*. Time becomes unavailable when the BER exceeds 10^{-3} in each of 10 consecutive seconds. Time remains unavailable until each of 10 consecutive seconds has a BER of less than 10^{-3}.

Performance objectives are then defined by the percentage of time that must be available, the percentage of available minutes that are degraded, the percentage of available seconds that are severely errored and the percentage of available seconds that are errored. Specifically, CCITT states the objectives for a *hypothetical reference connection* (*HRX*) which is defined to be a connection for a 64 kbits s^{-1} signal over 27 500 km. These objectives are:

fewer than 10% of minutes degraded

fewer than 0.2% of seconds severely errored

fewer than 8% of seconds errored (greater than 92% of seconds error-free)

The degradation builds up through a connection, so any one link in a network is allowed only a small contribution towards the overall objectives. The apportionment is in terms of the percentage time allowed with the various parameters. So, for example, a trunk link between two cities in the UK might be allowed:

fewer than 0.05% of minutes degraded

fewer than 0.005% of seconds severely errored

fewer than 0.05% of seconds errored

The error rates used in the definitions of the parameters are not subdivided: only the percentage time allowances.

For the purposes of assessing how much degradation is allowed in any given transmission system, the connection is assumed to consist of three grades of transmission: local, medium and high. These three roughly correspond to transmission through the local network, the trunk network and the international network. The local grade is assumed to be the 'worst' transmission, so is allowed the highest degradation. This hierarchy of transmission quality allows the final link to the end customer – where most of the transmission costs are incurred – to be as cheap as possible. Long distance transmission, which is shared among customers by multiplexing, (Chapter 8), can be cost effective even if its higher quality makes it more expensive.

3.7   Jitter and wander

The problem of jitter (irregularities in the timing of a digital waveform) was mentioned at the beginning of Chapter 2. It was implied that retiming of the waveform as part of regeneration can eliminate jitter. In practice, complete elimination of jitter is impossible – because the clock waveform used for the retiming will not itself be perfect. Consequently it is important to be able to quantify the amount of jitter on a digital waveform.

Fig. 3.24(b) shows a data waveform affected by jitter. The jitter appears as displacement of the actual timing interval boundaries from the nominal ideal boundaries. The timing interval boundaries are taken as the positive transitions of a clock waveform. Fig. 3.24(a) shows an *ideal* clock which provides the timing reference (you can think of this as the clock which originated the data – before it acquired jitter). The clock signal that would be derived from the *jittered* waveform is shown in Fig. 3.24(c). The jitter is then represented by the difference in time between the positive-going transitions of the reference clock and the positive-going transitions of the jittered clock, which is shown as the horizontal arrows in Fig. 3.24(d). Plotting this time difference against the nominal time, gives a series of points which may be considered to be samples of a continuous jitter waveform and which is shown in Fig. 3.24(e). Such a jitter waveform can be quantified as any other waveform can be: Fourier analysis can be used

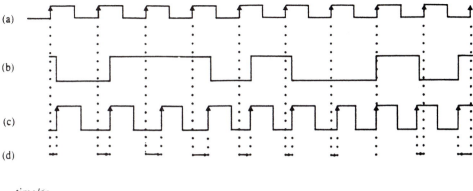

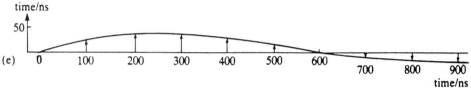

Fig. 3.24. Quantifying jitter.

to derive its spectrum, for example. A much simpler parameter, though, and one of particular practical importance, is the peak phase excursion resulting from jitter – that is, the difference between the peak negative excursion and the peak positive excursion, often known as the *peak to peak jitter*.

Instead of using time differences for the jitter measure, it can be normalised with respect to the clock period to give *unit intervals (UI)*. So an amplitude of 1 UI would be an excursion in time equal to the period of the clock.

Whether timing irregularities are viewed as jitter or wander depends upon the data signalling rate under consideration and the network characteristics. However, a typical boundary (for, say, a 64 kbits s⁻¹ signal) would be about 20 Hz. Wander is often thought of as a varying d.c. timing error, whereas jitter is an a.c. timing irregularity.

EXAMPLE

What is the peak jitter amplitude, measured in unit intervals, of the jitter on the waveform shown in Fig. 3.24?

Solution. The positive peak of the jitter (which occurs at around 200–300 ns) is about 40 ns. Since the clock period is 100 ns, this is $40/100 = 0.4$ unit intervals.

As is described in Chapter 5, it is possible to build circuits to reduce the amount of jitter on a digital waveform. However, low frequency jitter is particularly difficult to remove. For this reason it is often treated separately from higher frequency jitter and referred to as *wander*.

3.8   Summary

Commonly used statistical measures of noise (or other random variable) include the *mean* and *mean-square* values, the *variance*, and the *probability density function*. For the type of signal encountered in telecommunications, mean, mean square and variance can be defined either as time averages:

$$\bar{x} = \lim_{T_0 \to \infty} \frac{1}{T_o} \int_0^{T_o} x(t)\, dt$$

$$\overline{x^2} = \lim_{T_0 \to \infty} \frac{1}{T_o} \int_0^{T_o} x^2(t)\, dt$$

$$\sigma^2 = \lim_{T_0 \to \infty} \frac{1}{T_o} \int_0^{T_o} [x(t) - \bar{x}]^2\, dt$$

or in terms of the probability density function $p(x)$:

$$\bar{x} = \int_{-\infty}^{+\infty} x p(x)\, dx$$

$$\overline{x^2} = \int_{-\infty}^{+\infty} x^2 p(x)\, dx$$

$$\sigma^2 = \int_{-\infty}^{+\infty} (x - \bar{x})^2 p(x)\, dx$$

Mean, mean square and variance are related by the expression:
mean square = square of mean + variance

$$\overline{x^2} = (\bar{x})^2 + \sigma^2$$

which can be interpreted in terms of the power dissipated in a unit resistor:

total power = d.c. power + a.c. power

Independent random sources x and y with zero means satisfy the expression

$$\overline{xy} = \lim_{T_0 \to \infty} \frac{1}{T_0} \int_0^{T_0} x(t)y(t)\, dt = 0$$

and are said to be orthogonal. Hence for orthogonal variables

$$\overline{(x+y)^2} = \overline{x^2} + \overline{y^2}$$

and the powers of individual orthogonal components can be summed to give the total power.

The Gaussian distribution is an important model which closely approximates many sources of noise encountered in practice. For binary threshold detection in the presence of Gaussian noise, using voltage levels separated by V volts to represent the two digital states, the probability of error is given by:

$$P_e = Q(V/2\sigma)$$

where σ is the standard deviation of the noise (rms). Values of the Q-function are tabulated and plotted in standard reference works.

In the frequency-domain, noise (or other finite power signal) can be characterised by its power spectrum or spectral density. The total area under the power spectrum is a measure of the total power in the signal. For a linear system, input and output power spectra are related by the expression:

$$\text{output power spectrum} = \text{input power spectrum} \times |H(f)|^2$$

where $H(f)$ is the system frequency response function.

White noise has a constant spectral density for all frequencies of interest, and is often a useful model of thermal noise arising in a resistor and shot noise generated in active electronic components.

For the modelling of digital systems, a knowledge of the power spectra of random impulse trains is useful. The spectral density of a random, bipolar impulse sequence is a constant; that of a unipolar sequence also possesses discrete spectral lines at d.c. and all multiples of the sampling frequency.

A practical telecommunications link will be subject to noise other than Gaussian. To take this into account the CCITT has specified a number of parameters quantifying performance in terms of the percentages of minutes or seconds affected by errors. Jitter is a further source of signal degradation and ultimately errors.

Part 2
Processes

Suppose a digital signal at location A is to be transmitted to location B. In what way, if any, should the signal be processed before transmission over a particular digital link? At the very least the transmitted signal power needs to be sufficient for there to be an adequate signal to noise ratio at B, so amplification may be required prior to transmission. In practice there may be many other ways in which the signal is processed before transmission. Once at B, the conveyed signal needs again to be processed to recover the original signal and, possibly, to check that it has not been degraded in transmission.

In order to classify the various processes used for such purposes, we have found it convenient to use a layered model, akin to the OSI 7-layer model mentioned in Chapter 1. The OSI model itself is not, however, appropriate, since most of the processes discussed here would fall within the single 'physical' layer of the OSI model. Our model uses three layers and is shown in Fig. 2. It is assumed that the signal to be transmitted (which may originate at an inherently digital source such as a computer, or may be PCM coded speech, as discussed in Chapter 7) exists as a binary data sequence with an associated timing waveform. The three levels of processing for transmission are then as follows:

Channel coding is concerned with maintaining the integrity of the conveyed data sequence. Thus if the transmission channel is incapable of providing a low enough error rate for the needs of the data, the channel coding layer may incorporate error correction coding. The output from the channel coding layer to the line coding layer is another digital signal, consisting of a modified binary data signal with its timing waveform.

Line coding includes converting the sequences of binary digits into patterns suitable for transmission as pulses. It is responsible for maintaining the regular timing of the bit sequence (without transmitting a separate timing waveform), and may use error detection techniques to check if there are any faults on the line. Conceptually, the output from the line coding layer to the pulse layer may be viewed as a sequence of impulses. In practice, line coding and pulse generation will generally be performed together and the signal will never exist as a sequence of impulses. The distinction is nevertheless useful for analysing separate aspects of processing of a signal for transmission.

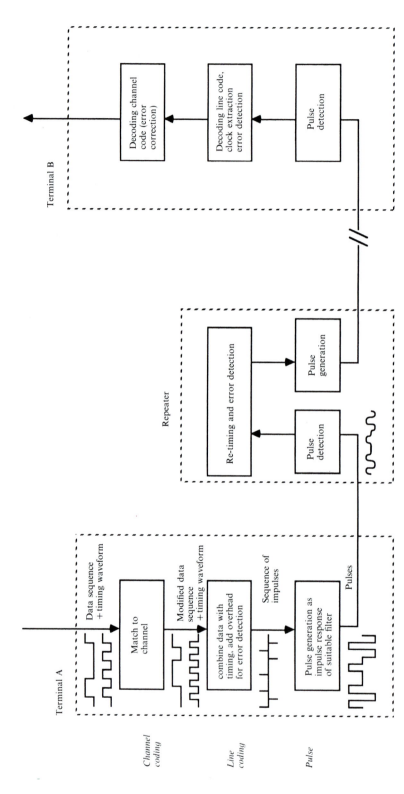

Fig. 2. A layered model of digital transmission.

Finally, *pulse generation* is concerned with transmitting and detecting individual pulses. Having modelled the input to this layer as a sequence of impulses, the pulse generator can be modelled as a filter with an impulse response equal to the desired pulse shape, as was discussed in Section 2.5. This is a sort of analogue layer, in that it is concerned with handling the voltage waveforms on the line. Methods to maximise the signal to noise ratio are a concern of this layer.

Notice in Fig. 2 that if the transmission incorporates a regenerative repeater, processing within the repeater will be at the pulse generation and line coding layers, but not at the channel coding layer. Channel coding is concerned with the complete end-to-end transmission.

The three chapters of this part adopt a 'bottom up' approach: starting with the generation and detection of pulses, followed by line coding and channel coding.

4

Pulses for digital transmission

4.1 Introduction

Armed with the general modelling framework developed in Part 1, and the layer model just presented in the Introduction to Part 2, we now turn specifically to the transmission of digital pulses. We begin with *baseband* digital signals – that is, with signals whose spectrum extends down to or near zero frequency. For much of the discussion the model of a transmission link shown in Fig. 4.1 will be used, in which a transmitted pulse is viewed as the impulse response of a suitable transmit filter $G_T(f)$. This idea was first introduced in Section 2.5. Although it does not correspond to practical methods of pulse generation, which (for baseband systems) would normally involve switching between appropriate voltage levels followed by appropriate filtering, the model is a useful conceptualisation, and enables different coding and pulse shaping techniques to be compared. The input data is represented by an appropriate sequence of impulses or delta functions, passed to the pulse layer from the line code layer. The channel itself has frequency response $C(f)$, and the signal processing prior to threshold detection at the receiver (filtering, equalisation) is combined into a single receive filter response $G_R(f)$.

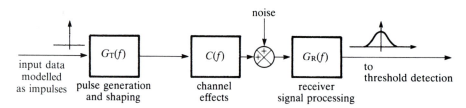

Fig. 4.1. Frequency domain representation of a digital telecommunication link.

4.2 Signalling rate and data rate

In Chapter 2 it was stated that the (theoretical) maximum rate of signalling over a bandlimited channel of bandwidth B Hz is $2B$ symbols per second; an illustration was given of how an alternating sequence of

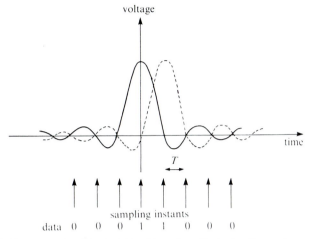

Fig. 4.2. Two successive, identical $(\sin x)/x$ pulses, illustrating the maximum rate of pulse transmission over a bandlimited channel.

states could be decoded provided the fundamental frequency could be passed by the channel. For the general case we turn to the model of Fig. 4.1, and suppose initially that the overall frequency response of the link $H(f) = G_T(f)C(f)G_R(f)$ can be made to approximate an ideal, distortionless, lowpass channel of bandwidth B. In other words, the received pulses at the threshold detector ideally have the characteristic $(\sin x)/x$ shape shown in Fig. 2.27 (or 2.28, assuming a suitable propagation delay) with zero crossings occurring every $1/T$ seconds, where $B = 1/2T$. Although such a characteristic cannot be realised perfectly, it can be approximated, as mentioned earlier.

Suppose now that we transmit symbols at a rate of one every T seconds (a signalling rate of $1/T$), with the receiver's threshold detector sampling the incoming signal also every T seconds. The overall received waveform can be found by superposition, as in Chapter 2. Fig. 4.2 gives the general idea, assuming unipolar signalling in which a binary 1 is represented by the transmission of a pulse, and a binary 0 by the absence of a pulse. Because of the location of the zero-crossings of the $(\sin x)/x$ waveform, a received pulse will contribute to the superposed, detected signal at the sampling instant corresponding to its own time slot only: at all other sampling instants it is identically zero. Exactly the same argument would apply to signalling using more than one pulse magnitude, as shown in Fig. 4.3(a) for two different positive levels and 4.3(b) for equal positive and negative levels (bipolar signalling).

Theoretically, then, by ensuring received pulses of $(\sin x)/x$ form, data can be transmitted unambiguously at a rate of $1/T$ symbols per second over a channel of bandwidth $1/2T$ Hz. In other words, a maximum

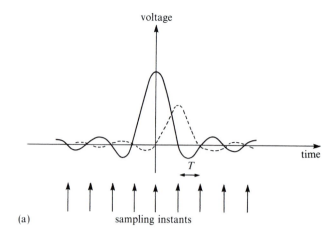

(a)   sampling instants

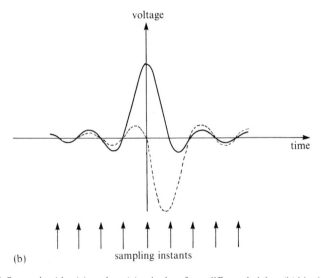

(b)   sampling instants

Fig. 4.3. Successive $(\sin x)/x$ pulses: (a) unipolar of two different heights, (b) bipolar of equal heights.

signalling rate of $2B$ symbols per second is possible over a channel of bandwidth B Hz. This theoretical maximum signalling rate is known as the *Nyquist rate*, after Harry Nyquist, the communications engineer who developed much of the theoretical background to pulse transmission in the 1920s (in the context of telegraphy).

4.2.1   Multi-level systems

In the case of a binary system, the signalling rate in *bauds* (the number of symbols per second) is equivalent to the data rate in bits per second. In non-binary digital systems, in which more than two different symbols are used, the relationship between signalling rate and data rate is different. Consider, for example, a quaternary baseband system using pulses of ± 1 V and ± 3 V as the four digital symbols. (These would be represented in the general model of Fig. 4.1 by impulses of four appropriate strengths.) $2B$ symbols per second can still be transmitted over a channel of bandwidth B Hz, but the signalling rate is no longer equal to the data rate as it was for the binary case. If four levels are used, each can be used to transmit two binary digits. For example

$+3$ V	11
$+1$ V	10
-1 V	01
-3 V	00

Here we tacitly assume that M is a power of 2. Similar ideas apply to codes where this is not the case, as will be discussed in Chapter 5.

For such a quaternary system using four distinct symbols, therefore, the data rate in bits per second is twice the signalling rate in bauds. Similarly, the data rate for an eight-symbol (8-ary) code is 3 times the signalling rate, while for a general 'M-ary' code using M discrete levels it is $\log_2 M$ times the signalling rate.

EXAMPLE

A digital transmission system uses 16 different signalling states, transmitting at 1200 baud. What is the maximum data rate in bits s^{-1}?

Solution. If sixteen states are available, each can represent 4 bits (0000 to 1111). The data rate is therefore four times the signalling rate $= 4800$ bits s^{-1}.

Note that there is an implied trade-off between the rate of transmission of information, channel bandwidth, and susceptibility to noise. We can increase the number of bits transmitted per second over a baseband system using a given bandwidth by increasing the number of discrete voltage levels employed as digital symbols. However, unless the transmitter power is increased (and there will be strict practical limitations on the extent to which this is possible), the voltage levels at the receiver will be correspondingly closer spaced, and it will be more likely that noise in the channel will result in an incorrect decision being made at the threshold detector.

4.3   Intersymbol interference (ISI)

In practice, trying to approach $(\sin x)/x$ pulses at a digital receiver would usually be counter-productive. Although such pulses are theoretically zero at sampling instants in all time slots other than their 'own', they have a considerable magnitude *between* these instants. The $(\sin x/x)$ function is highly oscillatory, and the side lobes take a comparatively long time to decay. So if the timing of the sampling at the receiver is less than perfectly regular, or the approximation to the perfect $(\sin x)/x$ function is insufficiently close, then the advantages of the approach may well disappear. If sampled even slightly out of phase, significant signal levels may be detected outside a pulse's own time slot: in such cases intersymbol interference (ISI) is said to occur. This is undesirable, since the combination of ISI, noise and other interference can result in an incorrect decision being made at the sampling instant about the digital symbol originally transmitted. Ideally, therefore, we would like to reduce the magnitude of the oscillations of the received pulse, while maintaining the desirable property of regular zero-crossings every T seconds.

Fig. 4.4 shows the pulse, introduced in Chapter 2, possessing 'a raised-cosine' spectrum. It has regular zero crossings, and for the pulse as drawn in the figure, $1/T$ symbols per second can again (theoretically) be transmitted without intersymbol interference. In this case, however, not only do the side-lobes die away much faster, but their heights are much lower and there are even additional zero-crossings midway between sampling instants. Furthermore, it is easier to design transmission systems approximating such a raised-cosine response, while maintaining reasonably linear phase.

The importance of linear phase to prevent the distortion of digital pulses was discussed in Section 2.2.

The combined effect of superposing a number of such (unipolar) pulses is illustrated in Fig. 4.5, which clearly shows their advantages for ISI. But note how the spectrum of such a pulse (shown in Fig. 4.4(b)) compares

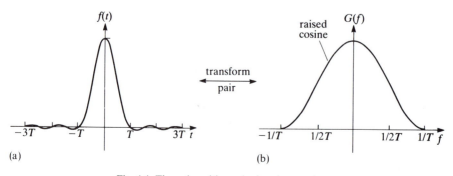

Fig. 4.4. The pulse with a raised cosine spectrum.

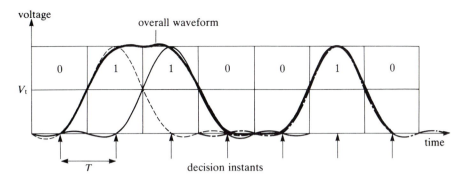

Fig. 4.5. Superimposing unipolar pulses with raised-cosine spectrum, illustrating theoretically zero intersymbol interference.

with that of a $(\sin x)/x$ function with the same spacing of zero-crossings: the raised-cosine spectrum has a bandwidth of $1/T$ Hz – double the earlier $1/2T$ Hz. Put another way, to use pulses with this raised-cosine spectrum, an overall bandwidth of B Hz allows signalling at only B baud – half the theoretical maximum. This is the price paid for the desirable ISI properties.

4.3.1 The raised-cosine family

The general technique of tailoring a desirable received pulse shape/spectrum to a given application is known as pulse shaping. There exists a 'family' of raised-cosine spectra and corresponding pulses, in which time-domain oscillations (and hence ISI characteristics) can be traded-off against spectral roll-off (and hence bandwidth requirements). Members of this family are usually characterised in terms of their 'percentage roll-off' in the frequency-domain, as illustrated in Fig. 4.6. The raised-cosine spectrum already introduced, which uses twice (100% more than) the theoretical minimum bandwidth, is said to have 100% roll-off; the 50% roll-off spectrum has a bandwidth 50% greater, and so on. The rectangular spectrum can therefore be considered a member of this family with zero roll-off.

Mathematical expressions for these spectra and pulses can be found in standard texts such as Schwartz, Peebles, and Bylanski & Ingram, so will not be reproduced here. The most important point is to note how relative ISI immunity – that is, a less oscillatory pulse – can be traded-off against bandwidth requirements. Note in passing, however, one additional feature of the 100% roll-off version: the time-domain pulse drops to half its maximum value exactly half a signalling interval beyond the peak. This

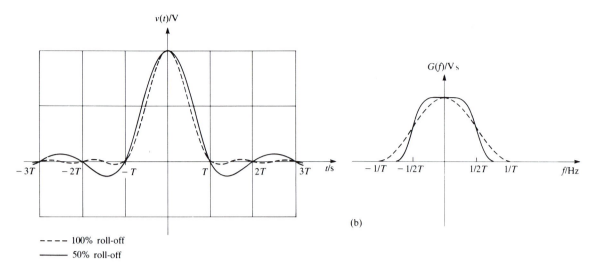

---- 100% roll-off
——— 50% roll-off

Fig. 4.6. Two members of the family of pulses with a raised-cosine spectrum.

has certain advantages in generating a clock waveform at the receiver, since threshold detection produces more regular transitions (for unipolar pulses) than with any other spectral roll-off.

The particular shape adopted in any given application will depend on various factors, such as the availability of bandwidth, the expected noise levels, the desired signalling rate, and so on. By way of illustration, the raised-cosine roll-off specified by the CCITT for a V.22 bis modem is 75%, while some satellite digital radio systems use roll-off factors of 30% or lower.

The characteristics of the received pulse will be determined by those of the transmitted pulse, the channel and the receiver. There may be little that can be done to modify the characteristics of an existing channel, but the designer often has the freedom to choose transmitter and receiver characteristics to provide the best compromise. From the point of view of ISI, appropriate pulse shaping can be provided at the transmitter, the receiver or a combination of the two. ISI is not the only problem, however; the effects of noise may well be as important or more so. The 'optimum' way of matching transmitter, channel and receiver in the presence of noise will be considered later.

4.3.2   Eye diagrams

A convenient practical way of assessing the level of intersymbol interference, as well as other signal degradations, is illustrated in Figs. 4.7 and 4.8.

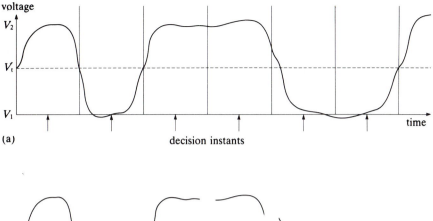

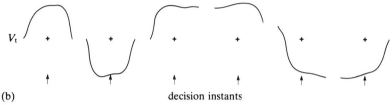

(a)   decision instants

(b)   decision instants

Fig. 4.7. Separating a binary signal into successive signalling elements.

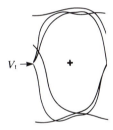

Fig. 4.8. Building up an eye diagram by superimposition.

The received, distorted waveform is separated into segments corresponding to individual signalling elements, which are then superimposed. A cross-hair has been added to part (b) of Fig. 4.7 to represent both the sampling instant and the binary threshold. A clear area around the cross-hair in the superimposed diagram of Fig. 4.8 is desirable, and implies a relatively low probability of misinterpreting a received symbol; in fact, direct numerical estimates of likely error rates can be made from the size of this clear area.

Because of its general appearance, such a representation is known as an eye diagram or a data eye. It can be obtained directly from a received baseband digital waveform by triggering an oscilloscope at the signalling rate from the clock pulses. Successive received symbols are then superimposed automatically on the oscilloscope screen. Fig. 4.9 is a

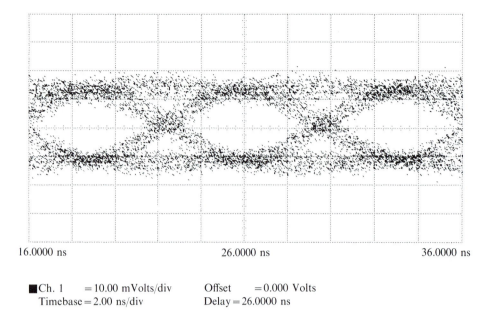

16.0000 ns   26.0000 ns   36.0000 ns

■Ch. 1   = 10.00 mVolts/div   Offset   = 0.000 Volts
   Timebase = 2.00 ns/div   Delay = 26.0000 ns

Trigger on External at Pos. Edge at −470.5 mVolts

Fig. 4.9. An eye diagram from a low-noise, practical system using approximately
raised-cosine spectrum pulses.

print-out of a display from a 140 M bit s^{-1} optical fibre transmission
system.

Three time intervals are visible, and individual pulses have been shaped
to approximate to a raised-cosine spectrum. Noise, ISI and jitter
contribute to the blurring of the trace.

4.3.3   Partial response signalling

An alternative way to combat intersymbol interference is, instead of
trying to eliminate it, to allow it to occur in a controlled manner. The
general principle has been known since the days of early telegraphy, when
the technique of 'doubling the dotting speed' was used to make better use
of available bandwidth.

Suppose that symbols are transmitted at a rate of one every T seconds
without intersymbol interference, and that a bipolar system is used such
that the received voltage levels are $+V$ for a binary 1 and $-V$ for a binary
0. Suppose further that an attempt is then made to double the
transmission speed, and that because of the finite rise time of the received
symbol, the expected voltage V cannot be reached within the new, shorter

signalling interval. The received pulse will spread into the next interval: substantial ISI occurs and individual symbols cannot be decoded accurately in the absence of further information. However, if we consider the transmitted symbols in pairs, it may be possible to make unambiguous decisions. If a half-width 1 is followed by another 1, or a half-width 0 by another 0, then the situation is equivalent to sending a single, full-width positive or negative pulse at the lower signalling speed. The received voltage will be $+V$ or $-V$. If, on the other hand, a half-width 1 is followed by a 0, or vice versa, then the two received pulses will tend to cancel out at the sampling instant. Hence we have the following decoding rules:

received voltage	transmitted symbol
$+V$	1 (previous symbol also 1)
$-V$	0 (previous symbol also 0)
near zero	complement of previous symbol

In this example, which is also known as a *duobinary* signalling scheme, it has been tacitly assumed that a pulse causes ISI only in the immediately following signalling interval. A conveniently-shaped pulse $g(t)$ with this specific property can be formed from a $(\sin x/x)$ waveform (as in Fig. 2.27) added to a replica delayed by one signalling period T. Such a pulse is shown in Fig. 4.10, together with its amplitude spectrum, $|G(f)|$, which can be shown to have a cosine (not a *raised-cosine*) shape. As can be seen, the pulse is completely restricted to a bandwidth of $1/(2T)$ Hz. If bipolar signalling is used, with $g(t)$ representing 1 and $-g(t)$ representing 0, then provided that the symbols are interpreted in pairs as described above, signalling can take place at the Nyquist rate of $1/T$ binary symbols per second. Note that the received pulse has zero value at all except two sampling instants, in contrast to the raised-cosine spectral shaping introduced earlier, where the received pulse was zero at *all* sampling instants outside its 'own' interval.

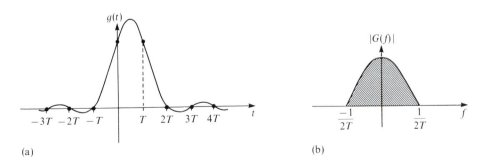

(a)   (b)

Fig. 4.10. A pulse for partial-response (duobinary) signalling, together with its amplitude spectrum.

EXERCISE 4.1

Make a rough sketch of the received waveform for the transmitted
sequence 110110, using bipolar pulses of the form of Fig. 4.10. Show
the approximate detected values of the received waveform at the
sampling instants. How are these values interpreted?

Various signalling schemes using this basic principle have been
proposed, and some have found practical application. One pulse which
has been used for telephony is shown in Fig. 4.11, together with its
amplitude spectrum. This scheme, known as *Lender's modified duobinary*
scheme, allows controlled ISI in the next symbol but one from the time
slot under consideration: the pulse consists of a $(\sin x/x)$ waveform from

> Note that duobinary
> signalling does not give
> us something for
> nothing! The signal at
> the receiver is
> interpreted in terms of
> three levels rather than
> two, and performance
> is thus more
> susceptible to noise.
> Duobinary signalling
> can therefore be viewed
> as another example of
> the trade-off between
> bandwidth and signal to
> noise ratio, as discussed
> earlier in the general
> context of multi-level
> codes.

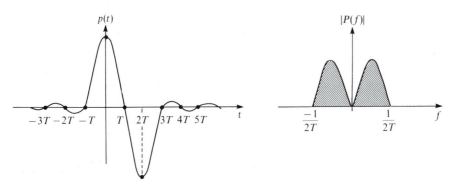

Fig. 4.11. The pulse used for Lender's modified duobinary scheme and its amplitude
spectrum.

which is subtracted a replica delayed by *two* sampling intervals. This
particular pulse shape possesses various advantages, including the
important practical one that its spectrum rolls-off to zero at low
frequencies, in contrast to the earlier duobinary spectrum. (The significance of this will be discussed in the next chapter.) Detailed consideration
of such *partial response signalling* is beyond the scope of this book, but the
topic will be taken up again briefly in Chapter 6.

4.3.4   Digital equalisation

Ideally, when sampled by the receiver at the signalling rate, an isolated
received pulse should give rise to a sample sequence of the form

$$\dots 0,0,0,1,0,0,\dots$$

Fig. 4.12. A received, distorted pulse suffering from ISI.

In other words, at only one sampling instant should the pulse take on a non-zero value (indicated as 1 here for convenience). Suppose, however, that the received, distorted pulse takes the form shown in Fig. 4.12, with non-zero sample values

$$0.30, 1.0, -0.20, 0.10$$

(The time origin of the received signal is arbitrary; it is shown coinciding with the first non-zero sample value.)

A discrete transversal filter of the type introduced in Section 2.6 can be used to process the incoming pulse so that the filter output takes on zero value at a given number of sampling instants either side of the maximum. Such a *zero-forcing equaliser* may be used in addition to other pulse shaping and filtering in transmitter and/or receiver. To illustrate the procedure, consider first a simple, three-term filter as shown in Fig. 4.13.

The filter unit sample response can be written down immediately as

$$h(n) = h_0, h_1, h_2$$

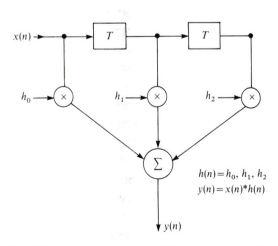

Fig. 4.13. A three-term transversal filter to counteract ISI.

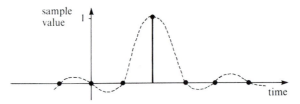

Fig. 4.14. The ideal response of the filter to the samples of Fig. 4.12.

Suppose that we aim to generate equalised sample outputs of the general form shown in Fig. 4.14 – that is, we accept an additional delay of one signalling (sampling) interval, and aim to force the output to zero at all sampling instants either side of a central peak of height 1. Using the convolution approach we can write

$$y_0 = x_0 h_0$$
$$y_1 = x_0 h_1 + x_1 h_0$$
$$y_2 = x_0 h_2 + x_1 h_1 + x_2 h_0$$

etc.

The design problem therefore becomes that of choosing suitable values of the filter coefficients.

EXAMPLE

The three-term filter of Fig. 4.13, with a three-term unit sample response, requires three coefficients to be specified. This in turn allows us to specify only three output values for a given input. Assuming that the desired output specification for the input distorted pulse is

$$y_1 = 0$$
$$y_2 = 1$$
$$y_3 = 0$$

that is, a central peak (delayed by one interval) with one zero each side, determine the required filter coefficients for the distorted pulse of Fig. 4.12.

Solution. From the results of convolution we can write

$$0 = x_0 h_1 + x_1 h_0$$
$$1 = x_0 h_2 + x_1 h_1 + x_2 h_0$$
$$0 = x_0 h_3 + x_1 h_2 + x_2 h_1 + x_3 h_0$$

Substituting in the input sample values, and bearing in mind that $h_3 = 0$ (unit sample response is $h_0, h_1, h_2, 0, 0, 0, \ldots$) leads to

$$0 = 0.3h_1 + h_0$$
$$1 = 0.3h_2 + h_1 - 0.2h_0$$
$$0 = h_2 - 0.2h_1 + 0.1h_0$$

Solving these three simultaneous equations gives

$$h_0 = -0.266$$
$$h_1 = 0.886$$
$$h_2 = 0.204$$

to three significant figures.

The effect of the transversal filter with these coefficients can be calculated using convolution. A spreadsheet is a convenient way of carrying out the computations quickly and easily, as illustrated in Appendix C. In this way the equaliser output sequence (to two significant figures) can be shown to be:

$$0.0, -0.08, 0.0, 1.0, 0.0, 0.05, 0.02, \ldots$$

This, illustrated in Fig. 4.15, is a distinct improvement over Fig. 4.12 from the ISI point of view. Notice, however, that the zero-forcing either side of the peak has resulted in the introduction of slight ISI at other sampling instants where there was originally none.

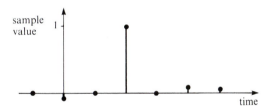

Fig. 4.15. The simulated response of the filter.

EXERCISE 4.2

(a) If you have access to a spreadsheet program, set up the spreadsheet for this example, shown in Appendix C, and investigate the sensitivity of the equaliser to slight numerical variations in coefficient values. How does the equaliser respond to input sequences other than that for which it was designed?
(b) Design a similar three-term equaliser for a pulse with the input sample sequence

$$\ldots 0, 0, 0, 0.2, 1.0, -0.1, 0.1, 0, 0, 0, \ldots$$

What additional ISI is introduced by the zero-forcing equaliser?

The zero-forcing approach can be generalised. The notation used above was chosen to correspond to the earlier discussion of discrete processing and convolution. The more usual notation for a zero-forcing equaliser is to denote the filter coefficients (and hence the unit sample response sequence) of a $2n+1$ term equaliser as

$$c_{-n}, c_{1-n}, \ldots c_{-1}, c_0, c_1, c_2, \ldots c_{n-1}, c_n$$

In such a general case, n zeros can be forced each side of a central peak delayed by n sample intervals; the filter coefficients are calculated from the $2n+1$ simultaneous equations expressed in matrix form as:

$$
\begin{pmatrix}
0 \\
0 \\
\cdot \\
\cdot \\
0 \\
1 \\
0 \\
\cdot \\
\cdot \\
0 \\
0
\end{pmatrix}
=
\begin{pmatrix}
p_0 & p_{-1} & \cdots & p_{-2n} \\
p_1 & p_0 & \cdots & p_{-2n+1} \\
\cdot & \cdot & \cdots & \cdot \\
\cdot & \cdot & \cdots & \cdot \\
p_{n-1} & p_{n-2} & \cdots & p_{-n-1} \\
p_n & p_{n-1} & \cdots & p_{-n} \\
p_{n+1} & p_n & \cdots & p_{-n+1} \\
\cdot & \cdot & \cdots & \cdot \\
\cdot & \cdot & \cdots & \cdot \\
p_{2n-1} & p_{2n-2} & \cdots & p_1 \\
p_{2n} & p_{2n-1} & \cdots & p_0
\end{pmatrix}
\begin{pmatrix}
c_{-n} \\
c_{-n+1} \\
\cdot \\
\cdot \\
c_{-1} \\
c_0 \\
c_1 \\
\cdot \\
\cdot \\
c_{n-1} \\
c_n
\end{pmatrix}
$$

where p_i represent the distorted pulse samples.

EXERCISE 4.3

Check that for $n=1$ this matrix equation leads to the coefficients of the three-term equaliser discussed earlier.

One of the advantages of using digital techniques is that the equaliser can be adjusted to changing channel characteristics. The filter coefficients are not set once and for all, but can be varied under microprocessor control. Coefficient setting is achieved by one of two methods. The first is to transmit a special *training sequence* before sending unknown data. Since the receiver 'knows' what training sequence to expect, the microprocessor can be programmed to optimise the equaliser coefficients so that the correct sequence is detected with the maximum eye opening. The training sequence should only have to be used again if the characteristics of the channel change for some reason.

Unfortunately, channel characteristics do sometimes fluctuate during transmission, and it is often desirable – particularly for systems transmitting data at high rates over less than ideal lines – to make adjustments as and when necessary, not just in response to a training sequence.

Microprocessor algorithms have been designed to monitor the data eye and continuously adjust equaliser coefficients so as to maximise the opening; this technique is known as *adaptive equalisation*.

4.4 Orthogonality and signal space

The material in this section and the next requires a highly mathematical treatment if it is to be made rigorous. This is widely available in other texts, and we have chosen here to concentrate on (i) presenting the important results without proof, and (ii) explaining their engineering significance.

Digital telecommunications involves the transmission of a relatively small number of distinct symbols. What is important is that the receiver should be able to distinguish clearly between these symbols even in the presence of noise, not that it should receive these symbols unchanged. A technique is therefore required to quantify the degree to which the discrete waveforms or digital symbols 'differ' from one another. To do this we introduce a new model of digital waveforms, based on the mathematical concept of orthogonality.

The following integral was introduced in Chapter 3 as a measure of the similarity or *correlation* between two finite-power signals $x(t)$ and $y(t)$:

$$\overline{xy} = \lim_{T_0 \to \infty} \frac{1}{T_0} \int_0^{T_0} x(t)y(t)\,\mathrm{d}t$$

Recall that if the result of this long-term average is zero the two signals $x(t)$ and $y(t)$ are said to be orthogonal, and their powers (mean-square values) add independently.

A similar expression can be used as a measure of the correlation between two finite-width pulses – or, indeed, between any two finite-energy signals over a given interval. Formally, two signals $s_1(t)$ and $s_2(t)$ are said to be orthogonal in the interval t_1 to t_2 if

$$\int_{t_1}^{t_2} s_1(t)s_2(t)\,\mathrm{d}t = 0$$

As a simple example of two orthogonal pulses, consider Fig. 4.16, which shows pulses restricted to the first and second half of the signalling interval respectively. (Such pulses could be used as symbols in a system where pulse *position* is modified in order to transmit the digital message.) The correlation integral must be zero over the complete signalling interval T (in this case two seconds) because the two pulses occupy completely separate portions of the time slot and their product is therefore zero over the time interval of interest. Given that the two pulses are completely separated in time it is perhaps not surprising that in a

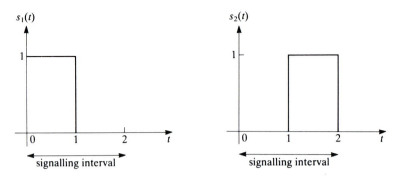

Fig. 4.16. Two orthogonal pulses.

well-defined mathematical sense they are 'independent' or 'completely different'.

EXAMPLE

Show that the signals $\sin \omega t$ and $\cos \omega t$ are orthogonal over any whole number of periods.

Solution. The proof follows most simply from the trigonometrical identity

$$\sin \omega t \cos \omega t = \tfrac{1}{2} \sin 2\omega t$$

The value of the integral of $\sin 2\omega t$ is clearly zero over any whole number of periods of the original sine and cosine functions, so the latter are orthogonal.

It may come as a surprise that two functions as apparently similar as sine and cosine should be orthogonal. The fact that they differ in phase by $\pi/2$ makes it possible to build electronic circuits which clearly distinguish between them, a highly practical consequence of their orthogonality.

Linear combinations of orthogonal signals can be represented as points in a suitable vector space. This idea is much less complicated than it sounds – a familiar example, widely used in electronics and telecommunications, is the humble phasor diagram. Because the sine and cosine functions are orthogonal, a sinusoid of arbitrary phase can be analysed into 'independent' sine and cosine (in-phase and quadrature) components. Fig. 4.17 is a reminder, showing how the sinusoid $5 \sin (\omega t + 37°)$ can be thought of as the sum of $4 \sin \omega t$ and $3 \cos \omega t$. The orthogonality of the sine and cosine components is reflected in the use of orthogonal (right-angled) axes. All linear combinations of the two base functions – expressions of the form $A \sin \omega t + B \cos \omega t$ where A and B are constants – can be represented by an appropriate phasor.

Note that the addition rule for orthogonal powers holds. The power of the composite signal $5 \sin (\omega t + 37°)$ of Fig. 4.17 is $5^2/2$ units $= 25/2$ units. Because the axes are at right-angles, Pythagoras's rule applies, and the

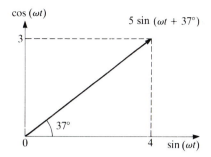

Fig. 4.17. A simple phasor diagram.

total power is equal to the sum of the powers in the orthogonal components: $25/2 = 9/2 + 16/2$. For non-orthogonal sinusoids (not at right-angles on the phasor diagram) this addition rule does not hold. For example, the terms $\sin \omega t$ and $\sin(\omega t + 180°)$ completely cancel out, giving zero total power even though each individual term has a power of 1 unit.

Analogous ideas apply to the representation of orthogonal pulses, rather than sinusoids. Fig. 4.18 is a two-dimensional diagram, known as *signal space*, representing all linear combinations of the orthogonal pulses of Fig. 4.16. The points at $+1$ on the horizontal and vertical axes represent the signals s_1 and s_2 respectively; similarly, point A represents the combination $s_1 + s_2$, while point B represents $s_2 - s_1$. The analogy with a phasor diagram is clear, although there are two important differences. The first is that it is usual to represent a signal by a single point in signal space (known as a *signal state*), rather than by a directed line as in a phasor diagram. The second difference is the way in which the distance of the signal state from the origin is interpreted; to understand this we need to note an important property of orthogonal pulses.

Consider the energy of a pulse composed of the sum of two signals s_1 and s_2 which are orthogonal over a signalling interval 0 to T and confined

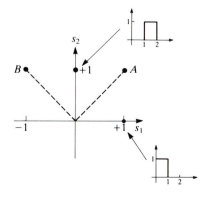

Fig. 4.18. The signal space representation of the pulses of Fig. 4.16.

completely to this interval (such as the pulses of Fig. 4.16). The energy of any pulse is found by integrating its instantaneous power over its duration. For any pulse s_1 confined to a signalling interval 0 to T, therefore:

$$\text{Energy} = \int_0^T s_1{}^2(t)\,dt$$

The energy of the composite pulse $s_1 + s_2$ is thus

$$\int_0^T (s_1 + s_2)^2(t)\,dt$$

$$= \int_0^T s_1{}^2(t)\,dt + \int_0^T s_2{}^2(t)\,dt + \int_0^T 2s_1(t)s_2(t)\,dt$$

Now, if s_1 and s_2 are orthogonal, the final term in the above expression is zero, and **the total energy is equal to the sum of the energies of the individual pulses**.

Returning to the signal space diagram of Fig. 4.18, we see that by interpreting the distance of a signal state from the origin as the *square root* of the pulse energy we have a completely consistent model. Adding the energies of two orthogonal signals to obtain the total energy then follows from Pythagoras's Rule, and becomes analogous to adding the powers of orthogonal phasor components to obtain the power of the resultant.

This is quite a subtle idea, and it may require a moment or two's thought. It should become clearer in the context of Fig. 4.18. The energy of a rectangular pulse of height V and duration T is $V^2 T$, so the energy of both s_1 and s_2 is 1 unit. Hence the square root of the energy is also unity and it was indeed valid to represent the two base signals by the two points at $+1$ on the horizontal and vertical axes respectively. A and B represent two new signals, $s_1 + s_2$ and $s_2 - s_1$, respectively; these are shown in Fig.

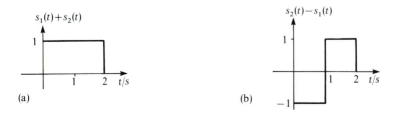

Fig. 4.19. Two new pulses generated by addition and subtraction of those in Fig. 4.16.

4.19. The distance of both A and B from the origin is $\sqrt{2}$, so the energy of each of the two new signals is 2 units. This value can be verified easily from the shape of the corresponding waveforms. Note also that because vectors A and B are orthogonal (at right-angles), so are the signals represented by these vectors.

EXERCISE 4.4

(a) Sketch the pulses represented by points X, Y and Z on Fig. 4.20. The axes represent the same orthogonal signals as in Fig. 4.18.
(b) Verify mathematically that the two signals A and B are orthogonal.
(c) Give the signal space position of a pulse orthogonal to pulse Z, and sketch its waveform.

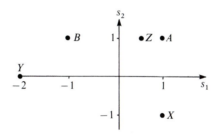

Fig. 4.20. Signal space diagram for Exercise 4.4.

4.5 Matched filter detection

The correlation expression, and the related notion of orthogonality, are useful tools for the analysis of the detection of digital signals in the presence of noise. The signals of interest are the discrete symbols which make up a digital waveform, and the time interval of interest is the duration of a signalling element, which will again be assumed to be from $0 < t < T$. Suppose that we use pulses which are orthogonal over the duration of a signalling element. The correlation integral then offers a route to making the best decision about which symbol is received in the presence of noise, provided that we can generate identical copies of the expected symbols at the receiver. We simply multiply the received signal in turn by each of the possible symbols and integrate over one sampling interval. In the absence of noise, the result will be zero if the locally generated pulse $p(t)$ is different from the received pulse, and

$$\int_0^T p^2(t)\,dt = \text{pulse energy}$$

if the locally generated pulse is identical to the received one. In the presence of noise or other imperfections we expect the results of the integration to differ somewhat from the ideal. Nevertheless, all except one of the integrations should be close to zero, thus allowing a clear decision to be made about which signal was transmitted.

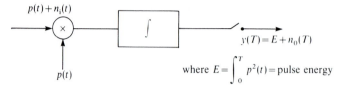

$p(t) + n_i(t)$

$y(T) = E + n_0(T)$

$p(t)$

where $E = \int_0^T p^2(t) = \text{pulse energy}$

Fig. 4.21. A correlation detector.

Fig. 4.21 shows such a *correlation detector*. Consider for a moment the effects of the correlation process on the signal and the noise. If the signal is what is expected, then multiplying it by a locally generated copy has the effect of squaring it, and thus giving a result which is always positive. Integrating this squared signal gives an output which continues to increase over the symbol period, reaching a maximum at time T. The noise, however, even when multiplied by the locally generated $p(t)$ will take on both positive and negative values, and so after integration its contribution $n_0(T)$ at the sampling instant will be comparatively low. It turns out that when the noise is additive, white, and Gaussian (AWGN) of zero mean then this process results in the best possible signal to noise ratio at the threshold detector at time T.

Such *optimal detection* can also be viewed as a filtering process. Consider a binary system using the presence or absence of a pulse $p(t)$ as the two symbols, and suppose that the correlation detector of Fig. 4.21 can be replaced by a filter – a filter with the special feature that its output after processing a complete pulse is equal to the energy of the pulse as with the correlation detector. The filter is a linear system, and hence can be modelled in the time-domain by its impulse response or, equivalently, in the frequency-domain by its frequency response. What must the impulse response $h(t)$ look like in order for the system output in response to $p(t)$ to be

$$\int_0^T p^2(t)\,dt$$

at time T? Recall that the system output is given by the convolution of $p(t)$ and $h(t)$, and that the convolution integral involves:

(i) the reversal of the impulse response $h(t)$;
(ii) multiplication by the input pulse waveform $p(t)$; and
(iii) integration of the result for all values of time shift.

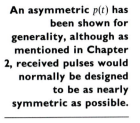
For a time shift equal to the symbol duration T, the time-reversed $h(t)$ therefore has to be identical to $p(t)$ – only then will $p^2(t)$ appear as the integrand. So the relationship between $p(t)$ and $h(t)$ must be as shown in Fig. 4.22. In mathematical notation $h(t)=p(T-t)$: reversing $h(t)$ and shifting it by an amount T gives a function identical to $p(t)$.

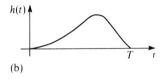

Fig. 4.22. The relation between pulse shape and matched filter impulse response.

EXERCISE 4.5

Check the result $h(t)=p(T-t)$ by substituting for $h(t)$ in the convolution integral, taking care with the dummy variable z.

Because of this relationship between $p(t)$ and $h(t)$, the impulse response of the filter is 'matched' in a very precise way to the expected pulse shape: the arrangement is known as a *matched filter*. Now, it can be shown from linear systems theory that if $h(t)=p(T-t)$, then (working in terms of angular frequency for convenience) $|H(\omega)|=|P(\omega)|$ and $\text{Arg}\,H(\omega)=-\omega T-\text{Arg}\,P(\omega)$. In other words, the matched filter amplitude frequency response is identical to the amplitude spectrum of the pulse. Such a filter emphasises those frequency components of the pulse which are most significant, de-emphasises those which are less important, and completely suppresses those which should not be present in the received pulse at all (but which may arise from the noise). The significance of the phase relationship is that the phases of the pulse frequency components are shifted such that the components all reinforce at time T, leading to a maximum output at the sampling instant. The noise components, on the other hand, have random phase angles and tend to cancel out (although not completely). As with the correlation detector, the result of all this is to give, for AWGN at the input, an optimum signal to noise ratio prior to threshold detection. In a practical realisation of a matched filter, additional scaling factors or gains will be involved in the receiver, such that $h(t) = kp(T-t)$ and $|H(\omega)| = k|P(\omega)|$, where k is a constant. Such gains will affect the noise and signal components identically, and do not affect the final S/N. For convenience, it will be assumed here that $k = 1$.

> Bear in mind that all this is merely a plausibility argument, not a proof! A full mathematical treatment can be found in, for example, Schwartz or Peebles.

4.5.1   Signal to noise ratio from an optimal detector

In the absence of noise, then, the output of an optimal detector at the sampling instant is numerically equal to the energy of the received pulse:

$$E = \int\limits_0^T p^2(t)\,\mathrm{d}t$$

However, to understand what this means for system performance (error rate) we also need to analyse the effect of the optimal detector on the input noise. A full mathematical analysis of matched filtering leads to the following remarkable result:

For AWGN with zero mean at the input, the noise at the filter output remains Gaussian with zero mean (no longer white) with a power (mean square value) numerically equal to $E\eta/2$, where E is again the energy of the pulse and η is the (single-sided) power density of the input noise.

> The *pulse* energy E appears in this expression for the output *noise* because the optimal detector has characteristics exactly tailored to the pulse – either by using a filter precisely matched to the pulse in the way outlined above or by correlating the pulse with a perfectly synchronised replica.

Since the output noise from the matched filter is Gaussian the Q-function can be used to find the error probability. At time T the wanted signal output is numerically equal to E, and the standard deviation (rms) of the output noise is $\sqrt{(E\eta/2)}$. Hence we have:

$$\begin{aligned}
P_e &= Q[\text{peak signal voltage}/2 \times \text{standard deviation of noise}]\\
&= Q[E/2\sqrt{(E\eta/2)}]\\
&= Q[\sqrt{(E/2\eta)}]
\end{aligned}$$

Rather than reproduce a general mathematical argument to prove this expression for an arbitrary pulse and its matched filter, the following example is included as an illustration of this important result.

EXAMPLE

The matched filter for a rectangular pulse has a rectangular impulse response. Its response to an incoming rectangular pulse is therefore

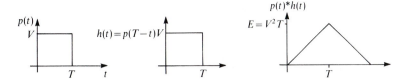

Fig. 4.23. Response of a filter matched to a rectangular pulse.

(by convolution) a triangular waveform with a maximum value of $E = V^2 T$, the pulse energy, at time T. This is shown in Fig. 4.23. (Such a matched filter can be implemented by integrating the incoming pulse as the charge on a capacitor up until time T, and then discharging the capacitor prior to the next pulse to prevent ISI. The circuit, effectively a correlation detector for a rectangular pulse, is known as an *integrate-and-dump* receiver.) Show that the output noise for AWGN at the input has a mean-square value $E/2\eta$.

Solution. The output noise will be coloured by the spectral characteristics of the matched filter – that is, the output mean-square noise voltage density will be given by $\eta\,|\,H(f)\,|^{\,2}$, as found in Chapter 3. If the impulse response is a rectangular pulse of height V and duration T, then from Chapter 2 the filter amplitude response $|\,H(f)\,| = VT\,|\,(\sin \pi fT)/\pi fT\,|$, and the output noise power spectrum (in single-sided form) will be as in Fig. 4.24. To calculate the total noise power we need to calculate the total area under this curve. Reference to standard mathematical tables reveals that

$$\int_0^\infty \frac{\sin^2 x}{x^2}\,\mathrm{d}x = \frac{\pi}{2}$$

If the appropriate change of variable is made, and the correct scaling factor applied, we find that

$$\int_0^\infty \eta V^2 T^2 \frac{\sin^2 \pi fT}{(\pi fT)^2}\,\mathrm{d}f = \frac{\eta V^2 T}{2}$$

The total area under the curve of Fig. 4.24 is therefore $\eta V^2 T/2$ (a value which agrees with a visual estimate). Since the energy of the rectangular pulse is $V^2 T$, then the total output noise power can be re-written as $E\eta/2$, the theoretical result quoted earlier.

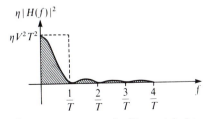

Fig. 4.24. Output noise power spectrum of a filter matched to a rectangular pulse.

The general expression $P_e = Q[\sqrt{(E/2\eta)}]$ for the error rate of an optimal detector is a highly significant one. It means that if a matched filter is used, then whatever the pulse shape, the error rate for AWGN is determined only by the pulse energy and the power density of the noise. The result is therefore applicable to systems other than baseband systems, as will be discussed below. The expression also sets a lower theoretical limit to the performance of a digital system in the presence of AWGN. It is not a result which can be used to calculate actual error rates in practical systems, because practical systems do not use perfect matched filters, and may be subject to noise other than Gaussian. Nevertheless it is extremely useful. It implies, for example, that an improvement in performance (lower error rate for a given noise density) cannot be obtained simply by changing the pulse shape, keeping its energy constant. There has to be some other specific improvement, such as a closer approximation to an optimum detector, or a reduction in intersymbol interference. Conversely, if the pulse shape needs to be changed for one of these specific purposes, then there is no theoretical reason why the error rate should suffer, providing the appropriate optimum detector can be realised and the pulse has the same energy.

The result is also useful for quantifying the deterioration in performance if the signalling rate over a noisy channel is increased. Using rectangular pulses of width T, for example, the pulse energy is V^2T. Doubling the signalling rate halves the pulse width, halving the pulse energy if the voltage remains the same. To maintain the same error rate, the signalling voltage would need to be increased by a factor of $\sqrt{2}$ to compensate. In practical cases, the penalty would usually be higher, as illustrated in the following example.

EXAMPLE

A copper coaxial system operating at 34 Mbaud uses 5 V rectangular pulses which are attenuated by 10 dB per km over 4 km, giving 50 mV pulses at the receiver. Suppose that the same cable is now to be used – still with 5 V transmitted pulses – at 136 Mbaud, when it will introduce attenuation of 20 dB per km. (The attenuation of copper coaxial cable increases strongly with frequency.) Estimate the maximum length of the new, higher-speed link, assuming that 5 V pulses are still transmitted and the error rate is to remain the same.

Solution. If the same error rate is to be achieved, then (assuming ideal receivers and AWGN) the same pulse energy must be maintained at the receiver. Since the signalling rate has been *increased* by a factor of 4, the pulse width has decreased by a factor of 4, and the received voltage must *increase* by $\sqrt{4} = 2$ to compensate – that is to 100 mV. To ensure 100 mV pulses at the receiver, however, the maximum permissible attenuation of the original 5 V pulses is $20 \log (5/0.1) = 34$ dB. At 20 dB per km, this limits the length of the new, higher speed link to 1.7 km, in contrast to the original 4 km.

One important feature of optimal detection should be emphasised. The integration of Fig. 4.21 can only be carried out accurately if (a) the precise pulse shape expected can be generated locally at the receiver and (b) perfect synchrony can be maintained between this pulse and the incoming signal. There are various ways of realising practical approximations to the theoretical optimal detector, but they all require this precise timing.

4.5.2   Signal space revisited

The signal space model can be related closely to the matched filter approach to signal detection. Recall that in signal space a pulse is represented by a vector of length $\sqrt{E}$, where E is the energy of the pulse. For the type of 'on-off' system considered so far, in which the two symbols are a pulse $p(t)$ and its absence, only one dimension is required in signal space: the two symbols are represented by points at the origin (absence of pulse) and at $+\sqrt{E}$ (presence of pulse). Now, the effect of noise is to make the received pulse less like the intended pulse. This can be interpreted in signal space as shifting the precise position of the point representing the received symbol. For AWGN, in fact, we can imagine this uncertainty as a Gaussian distribution centred around the ideal signal space location, as illustrated in Fig. 4.25. A perfectly received $p(t)$ in the absence of noise will be interpreted by an optimum detector as a signal located at precisely $\sqrt{E}$. Every other received waveform will be allocated to some other point along the line, and the decision whether 0 or 1 was transmitted will be

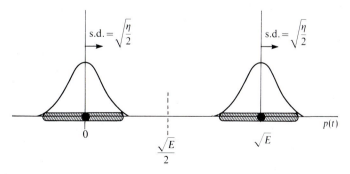

Fig. 4.25. The effect of additive, white, Gaussian noise in signal space.

made on the basis of whether this point is closer to zero or $\sqrt{E}$. We can imagine a signal space decision threshold at $\sqrt{E}/2$, so that an error will be made if noise displaces the binary 0 signal state by more than $+\sqrt{E}/2$ or the binary 1 state by more than $-\sqrt{E}/2$.

Indeed, we can go further. For simple threshold detection of a unipolar binary signal the probability of error was given by the expression

$$P_e = Q(\text{peak signal level}/2 \times \text{noise standard deviation})$$

For optimal detection we have

$$P_e = Q[\sqrt{(E/2\eta)}]$$

Now, however, we are working with a Gaussian distribution of signal states on a root energy scale. For 'threshold detection' in signal space, $\sqrt{E}$ corresponds to the earlier 'peak signal level' and $\sqrt{(2\eta)}$ to 'twice the standard deviation'. The standard deviation of the Gaussian distribution *in signal space* is therefore given by

$$\sigma_{\text{sig}} = [\sqrt{(2\eta)}]/2 = \sqrt{(\eta/2)}$$

EXAMPLE

The binary signalling technique in which the two symbols used are a pulse $p(t)$ and its inverse $-p(t)$ is sometimes known by the generic term 'antipodal signalling'. (Bipolar NRZ and RZ are simple examples.) Compare ideal error rates for antipodal and on–off signalling, assuming optimum detection in the presence of AWGN.

Solution. Antipodal signalling again needs only one-dimension in signal space, since only one basic pulse shape is used. Assume (for reasons which will become clear in a moment) that the two pulses $\pm p(t)$ are represented by the locations $\pm\sqrt{E}/2$ in signal space, as shown in Fig. 4.26. With a 'threshold' this time at 0, an error will again occur if noise displaces the one state by $+\sqrt{E}/2$ in signalling space or the other by $-\sqrt{E}/2$: the probability of error is therefore exactly as for the on–off case, $P_e = Q[\sqrt{(E/2\eta)}]$. The crucial point to note is that the error probability is identical because the distance between the states in signal space is identical: $\sqrt{E}$ for both the on–off and antipodal cases. The same error probability would be obtained with any two states separated by $\sqrt{E}$ in signal space, assuming the correct matched filter(s) for the particular pulse shape(s).

Note that the energy of each antipodal pulse is $E/4$, as opposed to E for the on–off pulse. Antipodal signalling therefore requires only half the average power of on–off signalling to achieve the same error rate, assuming equal numbers of binary ones and zeros. (Twice as many pulses will be sent, but each has only one quarter of the energy.) The

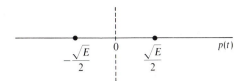

Fig. 4.26. Signal space for antipodal signalling.

situation is exactly analogous to the earlier comparison of unipolar and bipolar binary signalling using full-width rectangular pulses, but it has now been generalised to any pulse shape, assuming AWGN and optimal filtering.

To sum up, the advantages of the signal space representation are:

– any shape of pulse can be modelled
–the spreading of the point in signal space is a function of the (AWGN) noise power density alone
– the distance between points in signal space is a direct measure of the probability of one state being misinterpreted as another

4.5.3   Practical baseband receivers

This discussion of matched filtering and optimal detection has so far completely neglected intersymbol interference (ISI). Yet for many systems ISI can be as troublesome as channel noise. In some cases it may be possible to use the pulse-shaping approach of Section 4.3 while at the same time approximating an optimal detector. Suppose that we want pulses with a spectrum approximating to a suitable raised-cosine at the threshold detector. Assuming first that the channel has been engineered to have a flat, non-distorting, frequency response over the bandwidth of interest, then the received pulse (using the notation of Fig. 4.1) will have a spectrum of the form $G_T(f)G_R(f)$, since $C(f)$ is assumed to be unity. For optimal detection, the receiver filter $G_R(f)$ is to be matched to the pulse with spectrum $G_T(f)$, so that $|G_T(f)| = |G_R(f)|$. So if the overall received pulse spectrum $G_T(f)G_R(f)$ is to be of raised-cosine form, then

$$|G_T(f)| = |G_R(f)| = \sqrt{(\text{raised-cosine})}$$

The phase characteristic would also be designed to be a realisable approximation to the matched filter requirement.

Many real systems cannot be assumed to have flat channel attenuation over the bandwidth of the transmitted pulse, however. Coaxial cable, for example, has an attenuation proportional to the square root of frequency. It is possible to compensate for this by adding an equaliser to 'undo' the channel attenuation – in the coaxial case, the equaliser would have to have a *gain* proportional to the square root of the frequency. The problem now is that the equaliser also boosts high-frequency noise, which is therefore no longer white (even if it was in the first place). To reduce the effect of this additional high-frequency noise, it is desirable for the receiver filter to attenuate more at higher frequencies than is implied by the expression

$$|G_R(f)| = \sqrt{(\text{raised-cosine})}$$

derived above.

One common approach in practice when transmitting over copper cables is to concentrate more energy into the higher frequency components of the transmitted pulses. To put it another way, the equaliser boosts the high-frequency noise, so we boost the higher frequencies of the transmitted signal, and then attenuate both at the receiver to improve the total signal to noise ratio. A typical approach is to transmit half-width RZ (return to zero) pulses with their stronger higher frequency components, and to design the receiver filter such that the overall effect (including the equaliser) approximates to a suitable raised-cosine spectrum. A similar technique is used in analogue frequency modulation (known as pre- and de-emphasis), and in Dolby noise reduction on audio cassettes.

A further consideration in such systems is crosstalk arising from capacitative or inductive pick-up between physically adjacent metallic conductors. Such crosstalk tends to worsen at higher frequencies, since electromagnetic radiation from the conductors occurs more easily. A quantitative analysis of all the factors will in general be complex, and often involve computer simulation to suggest some 'best' pulse shape, equaliser and receiver filter to give the minimum error rate.

> In practice the equaliser and receiver filter are often combined into one electronic circuit. Some systems might incorporate an additional zero-forcing equaliser of the type described in Section 4.3.

4.6 Sinusoidal pulses

All the pulses considered so far have been baseband – that is, their spectra extend down to, or near, zero frequencey. Such pulses are often used for transmission over short distances using copper cable of one sort or another (in local networks, for example). However, for digital transmission over a radio link, or using a bandpass telephone channel originally designed for analogue speech, pulses with rather different spectral characteristics are required. Specifically, the pulse spectrum must be concentrated within a particular, *bandpass*, range of frequencies – either to match the limited bandwidth allocated, or as a result of the physical limitations of the transmission medium. For such purposes sinusoidal pulses are used as the digital symbols.

> The use of sinusoidal signalling elements for digital transmission was first mentioned in Chapter 1.

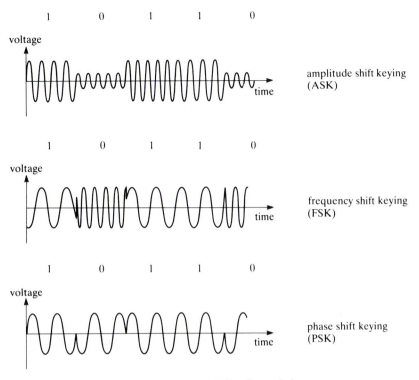

Fig. 4.27. Some sinusoidal signalling techniques.

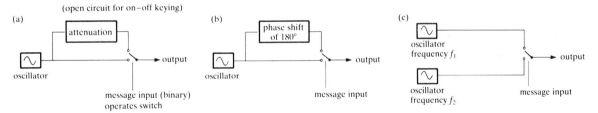

Fig. 4.28. Generation of sinusoidal pulses.

Fig. 4.27 illustrates some of the various ways of using such sinusoidal pulses. In *amplitude shift keying* (*ASK*) sinusoidal pulses of different amplitude are employed as the different symbols. (In the particularly simple case of *on–off keying* (*OOK*) the binary symbols are transmitted as the presence or absence of a sinusoidal pulse.) In *frequency shift keying* (*FSK*) sinusoidal pulses of different frequency are used; in *phase shift keying* (*PSK*) it is the phase of the sinusoidal pulse which distinguishes the transmitted symbols. The generation of such digital signals is illustrated conceptually in Fig. 4.28. To gain an insight into the spectral characteristics of such signals, let us first consider the elementary sinusoidal pulse shown in Fig. 4.29.

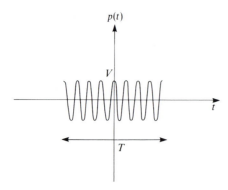

Fig. 4.29. An elementary sinusoidal pulse consisting of a whole number of cycles.

4.6.1   Spectra of sinusoidal pulses and pulse trains

The spectrum of the pulse of Fig. 4.29 can be derived in a number of ways. The most direct is to compute the appropriate Fourier integral (see Section 2.4.1 and Appendix A). Here, however, an alternative method based on convolution will be presented, which it is hoped will give an additional appreciation of this particular technique. Recall from Section 2.7.2 that convolution in the time-domain is equivalent to multiplication in the frequency-domain. It can be shown that the converse is also true: namely, multiplication in the time-domain is equivalent to convolution in the frequency-domain. In symbolic notation, if

$$p(t) = g(t)v(t)$$

then

$$P(f) = G(f) * V(f)$$

where, as before, the symbol * denotes the convolution operation

$$G(f) * V(f) = \int_{-\infty}^{+\infty} G(z)V(f-z)\,dz$$

Note that z is again a dummy variable, but now represents shifts in frequency rather than time.

The evaluation of frequency-domain convolution integrals need not concern us here. In general, functions like $G(f)$ and $V(f)$ take on complex values, and the mathematical operations involved in carrying out the integration may be quite complicated. The particular case of interest here, however, is reasonably straightforward, and the graphical technique introduced in Section 2.7.2 can be applied.

The pulse of Fig. 4.29 can be viewed as the result of multiplying together a rectangular pulse of unit height and duration T centred on the time origin, and a cosine function of amplitude V: the rectangular pulse has the effect of 'gating' or 'windowing' a section of the continuous sinusoid, as shown in Fig. 4.30. The spectrum of the sinusoidal pulse is therefore the convolution of the spectra of these two time-domain functions, as illustrated in Fig. 4.31. The spectrum of the rectangular pulse has the familiar $(\sin x)/x$ form, while that of the cosine function consists of lines of height $V/2$ at frequencies $\pm f_0$. Although we shall not prove it here, a spectral line can be represented mathematically by a frequency-domain delta function or impulse with a strength equal to the amplitude of the frequency component. The spectrum of the windowed sinusoidal pulse is therefore found by convolving the $(\sin x)/x$ function with the delta functions of strength or area $V/2$ at $\pm f_0$.

Recall that the graphical method of convolution involves reversing one function, sliding it over the other, multiplying, and *taking the area of the resulting new function* for all possible shifts. The $(\sin x)/x$ function is symmetric, so we can forget the reversal. Let us begin by shifting it a 'distance' f_0, which will lead to the value of the spectrum of the composite pulse at a frequency f_0. After such a shift the central peak of the $(\sin x)/x$

Fig. 4.30. The sinusoidal pulse of Fig. 4.29 as a windowed sinusoid.

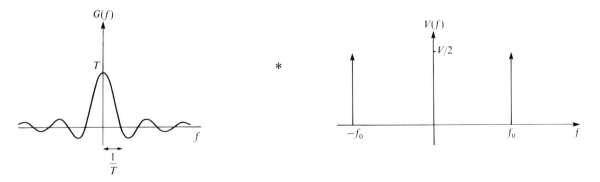

Fig. 4.31. Windowing in the frequency domain.

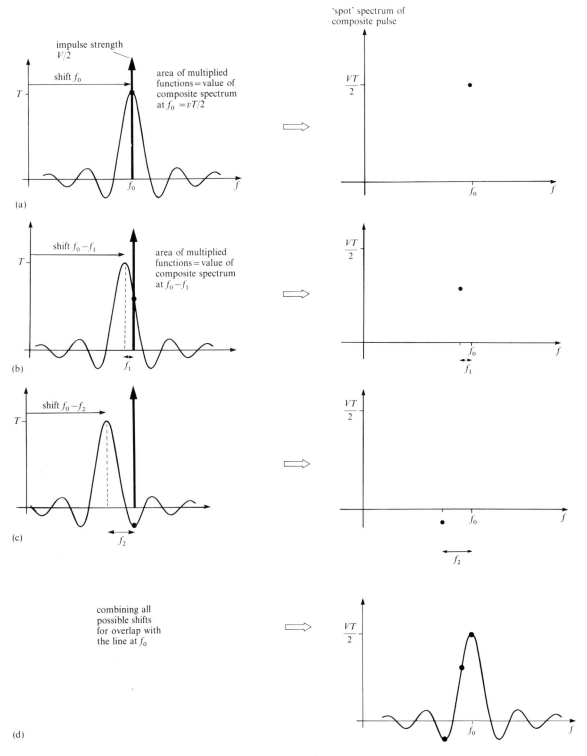

Fig. 4.32. Building up the spectrum of the windowed sinusoid by frequency-domain convolution.

function overlaps one of the delta functions (spectral lines). This is shown in part (a) of Fig. 4.32. To obtain the result of the convolution integral for this shift, and hence the composite pulse spectrum at the 'spot' frequency f_0, we multiply the two functions and find the *area* of the result. The result of the multiplication is zero except where the delta function and the $(\sin x)/x$ function coincide; at that point the result is simply a new delta function, with strength or *area* $V/2$ multiplied by the local value of the $(\sin x)/x$ function – in this case, therefore, $(V/2) \times T = VT/2$. This area $VT/2$ is therefore the value of the spectrum of the composite pulse at the frequency f_0. Shifts other than f_0 give other spot spectral values, as illustrated in parts (b) and (c) of Fig. 4.32. Repeating the process for other appropriate frequency shifts (Fig. 4.32(d)) we see that the overall result of the frequency-domain convolution is to reproduce a suitably scaled version of the $(\sin x)/x$ function about the frequency f_0. A similar effect occurs as a result of the second delta function at $-f_0$, leading to the complete spectrum of the composite sinusoidal pulse shown in Fig. 4.33. The general rule is quite simple: convolving an arbitrary function $f(x)$ with a delta function at position x_0 just shifts the original function through a distance x_0, scaling it by the strength of the delta function.

Appendix A gives a formal statement of this result as the 'modulation' property of the Fourier transform.

A number of important conclusions can be drawn from the preceding discussion. The first is to note that the sinusoidal pulse, like the square pulse used as its 'window', has a theoretically infinite bandwidth: the side lobes in the amplitude spectrum fall off inversely proportionally to the frequency displacement from the central peak, but remain significant even at substantial distances away. So to ensure that the signal is concentrated within a given bandpass spectral region, a more appropriate shaping than a simple rectangular window will often need to be applied to the sinusoidal segment – just as rectangular pulses are often inadequate for

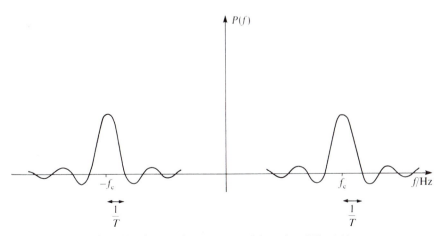

Fig. 4.33. The complete spectrum of the pulse of Fig. 4.29.

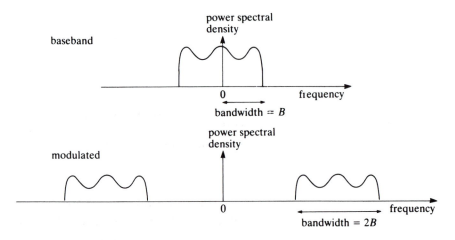

Fig. 4.34. Frequency translation by multiplication with a sinusoid doubles the signal bandwidth.

baseband signalling. The second important point is that, although neither the baseband window spectrum nor the bandpass sinusoidal pulse spectrum has a clearly defined total bandwidth, in a real sense the bandwidth of the latter is twice that of the former. Fig. 4.34 illustrates this in general terms for any baseband signal of bandwidth B frequency-shifted by multiplying it with a sinusoid. Because both the positive and negative frequency components of the baseband signal are shifted by the convolution process in both positive and negative directions, the bandpass signal extends *both above and below* its central frequency by B, giving an overall bandwidth of $2B$.

Let us now compare some of the important spectral characteristics of the various types of digital signalling using sinusoidal pulses. To do this, we can again use the pulse generation model of Fig. 4.1, in which an impulse train is applied to a suitable filter. In practice, waveforms would not be generated in this way, but by switching or shaping a sinusoid as shown in Fig. 4.35 for ASK – a baseband digital signal can be said to modulate a carrier sinusoid.

Strictly speaking, switching a carrier with a baseband pulse train is only *exactly* equivalent to generating individual sinusoidal pulses like $p(t)$ in Fig. 4.29 when the baseband pulse train is synchronised to the carrier such that f_c is an integral multiple of $1/T$. Only then will there be an integral number of carrier cycles in each signalling interval, so that successive switched carrier symbols are identical, rather than in some phase relationship defined by the ratio of f_c to T. Such synchronisation is sometimes employed in practical systems (by deriving both data timing and carrier from a single clock), sometimes not. (For the level of treatment

Many of the standard results of analogue modulation theory can therefore be carried over directly to digital transmission using sinusoidal pulses.

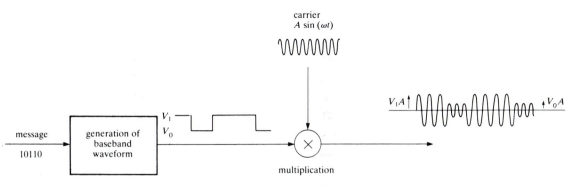

Fig. 4.35. Modulating a sinusoid with a baseband digital signal.

in this chapter we do not need to go into precise details, although a number of important consequences of synchronisation or the lack of it will be discussed later.) In any event, the general model of Fig. 4.1 remains a highly useful one which reflects the 'layer' analysis presented earlier and also brings a conceptual unity to baseband and sinusoidal digital signals. To generate our sinusoidal pulses, therefore, we imagine a filter with the frequency response $P(f)$ of Fig. 4.33 generating the pulse $p(t)$ of Fig. 4.29 as its unit impulse response.

Consider first the cases where the inputs to the pulse generator are the random binary bipolar and unipolar impulse trains shown in the previous chapter as Figs. 3.15 and 3.16 respectively. In the bipolar case the filter output will be a random sequence of $p(t)$ or its inverse $-p(t)$. In other words, the two possible symbols differ in phase by 180 and the filter output is a binary PSK waveform similar to that illustrated earlier in Fig. 4.27(c). The power spectrum of this random PSK waveform can be derived by the method of Section 3.3.3. Not surprisingly, it turns out to be a frequency-shifted version of the baseband, *bipolar* NRZ power spectrum of Fig. 3.17(a), as shown in Fig. 4.36(a). A filter input consisting of a *unipolar* impulse train, on the other hand, will generate an OOK waveform. The power spectrum in this case is shown in Fig. 4.36(b): it is equivalent to a frequency shifted version of the *unipolar* NRZ power spectrum of Fig. 3.17(b), and includes a spectral line at the carrier frequency corresponding to the shifted d.c. component of the baseband spectrum.

Remember that in both OOK and binary PSK the bandwidth requirement is double that of the corresponding baseband signal. As a broad rule of thumb we can assume that a binary baseband waveform requires a bandwidth approximately equal to the signalling rate (this value corresponds to 100% raised-cosine spectral shaping of a NRZ waveform). Using this estimate of the baseband bandwidth, both OOK

It can be shown that *m*-ary PSK occupies the same bandwidth as would an ASK signal modulated using the same (*m*-level) baseband signal, so the bandwidth of PSK in general may also be taken to be approximately twice the signalling rate. Of course, if more than two ASK amplitudes are used, or more than two phases, then each can represent more than one binary digit, and the data rate is correspondingly higher than the signalling rate, as discussed in Section 4.2.

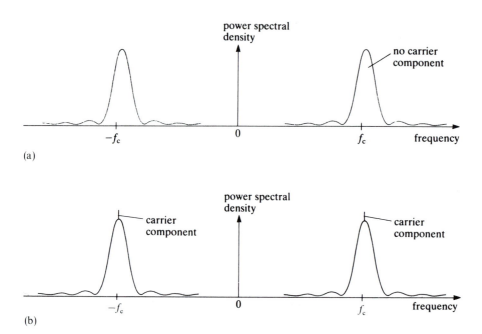

Fig. 4.36. Power spectrum of random binary (a) PSK and (b) OOK waveforms. Note the spectral line at the carrier frequency in (b).

and binary PSK – and, indeed, ASK and PSK in general – requires a bandwidth equal to *twice* the signalling rate.

The case of FSK is rather more complicated, and only the broad features will be mentioned here. One approach is to consider binary FSK as two interleaved OOK signals using two different frequencies. The power spectrum of an FSK signal might therefore be expected to be the sum of two OOK spectra, implying a double peak and an overall bandwidth requirement of $f_1 - f_0 + 2B$ as shown in Fig. 4.37. This is valid providing the two frequencies are sufficiently widely separated, but in practice the use of such widely separated frequencies might well be wasteful of scarce bandwidth resources. There is also another complication. Implicit in the approach just presented to FSK, and shown explicitly in Fig. 4.27(b) earlier, is the existence of a discontinuity in phase at each frequency transition. An alternative approach to FSK is to switch a single oscillator between frequencies in such a way that there is no phase discontinuity, as shown in Fig. 4.38. Such *continuous-phase* FSK (CPFSK) has rather different spectral characteristics. Fig. 4.39 shows an example: the (single-sided) power spectrum of random binary data coded using the particular type of CPFSK known as *minimum shift keying* (MSK). In MSK the two frequencies used have the minimum theoretical

This theoretical minimum frequency separation for distinguishing between the two digital states turns out to be $B/2$ for a signalling rate of B. In fact, theoretical optimal performance in the presence of noise is achieved with a spacing slightly higher than this at $0.7B$, and a spacing of about this value is often used in practice. The mathematical analysis of MSK (and FSK in general) is highly complex, and will not be considered any further here.

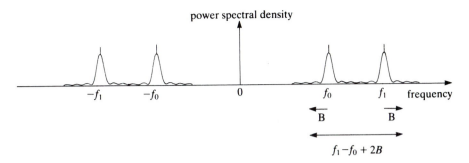

Fig. 4.37. Power spectrum of a random, binary, FSK waveform.

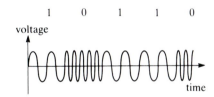

Fig. 4.38. A continuous phase FSK waveform.

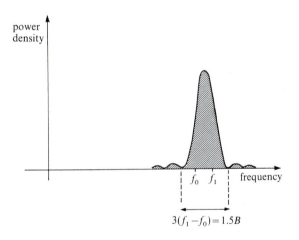

Fig. 4.39. Power spectrum of a random, binary MSK waveform.

separation: this fact, together with the absence of phase discontinuities at frequency transitions, results in a power spectrum with a single central peak, no discrete components, and a very fast spectral roll-off away from this peak.

4.6.2   Sinusoidal pulses and signal space

The signal space approach introduced in Section 4.4 can be applied
directly to sinusoidal pulses. As before, the individual pulses are plotted as
points in an appropriate signal space, the distance of a point from the
origin corresponding to the square root of the pulse energy. OOK and
ASK use only one basic pulse shape, so require only one dimension in
signal space. For OOK, therefore we have Fig. 4.40(a), with points at the
origin and $\sqrt{E}$, where E is the pulse energy; multi-level ASK would
include other points at positions on the line corresponding to the square
roots of the individual pulse energies.

   Binary PSK using phases of $0°$ and $180°$ is very similar. In this case the
two symbols are the inverses of one another. One dimension is still
sufficient, therefore, the two symbols being represented by the points
$\pm\sqrt{E}$ as shown in Fig. 4.40(b). Note that Figs. 4.40 (a) and (b) are
identical to those for unipolar and bipolar baseband binary signalling:
only the pulse shapes differ.

Fig. 4.40. Signal space representation of (a) OOK and (b) binary PSK.

With PSK using phases other than $180°$ apart, however, one-
dimensional signalling space will no longer do, since the various pulses no
longer differ simply by a scaling factor. Consider, for example, 4-level
(quaternary) PSK using phases $0°, 90°, 180°, 270°$, where each symbol has
energy E. In this case a two-dimensional signalling space is appropriate,
whose orthogonal axes represent windowed $\sin \omega t$ and $\cos \omega t$ pulses
respectively. The *signal constellation* for such QPSK is shown in Fig. 4.41.

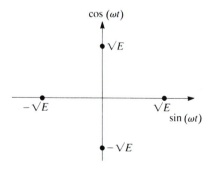

Fig. 4.41. Signal space for 4-phase PSK.

EXERCISE 4.6

Sketch the signal constellation for 8-level PSK using the eight equally-spaced phases 0°, 45°, 90°, 135°, etc. ... using the same orthogonal axes as Fig. 4.41.

It is also possible to have schemes which use sinusoidal pulses which differ in both amplitude and phase. For example, Fig. 4.42(a) shows the signal space constellation for *quadrature amplitude modulation (QAM)* using the four phases and two amplitudes illustrated in Fig. 4.42(b).

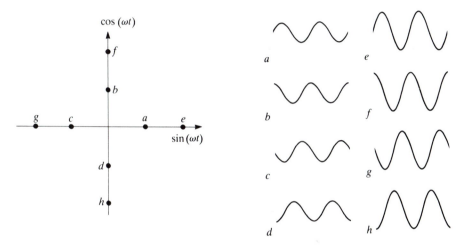

Fig. 4.42. Eight state QAM: (a) signal constellation and (b) signalling elements.

EXERCISE 4.7

Sketch the signal constellation of a QAM scheme using eight states, one of energy 1 and one of energy 9 arbitrary units at each of the phases 45°, 135°, 225°, and 315°.

To represent FSK in signal space we need orthogonal axes representing sinusoidal pulses at different frequencies. So, for example, Fig. 4.43 shows the signal space constellation of FSK using 200 Hz and 300 Hz to represent binary 1 and 0 respectively. For FSK using more than two states (frequencies), more than two orthogonal axes are required. Fig. 4.44 shows the principle for ternary FSK; for *m*-ary FSK where *m* > 3 it becomes impossible to draw the signal space, although the concept is still valid and useful for mathematical analysis.

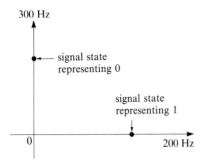

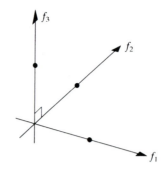

Fig. 4.43. Signal states of binary FSK.   Fig. 4.44. Ternary FSK in signal space.

4.6.3   Noise

The effect of noise can be modelled in ASK, PSK and FSK signal space in much the same way as was done in Section 4.5.2 for baseband signalling. There, noise was viewed as displacing the position representing the affected pulse in (one-dimensional) signal space. The same is true here, but points can now be displaced in more than one dimension, as illustrated in Fig. 4.45 for QPSK. Remember that noise modelling using signal space diagrams assumes a signal set made up of combinations of orthogonal base signals, and optimal (matched filter) detection. In effect, the receiver checks how well the received, noisy, pulses match what is expected, and 'allocates' them to a particular position in the appropriate signal space. The effect of noise is to shift the detected position in signal space away from the ideal. In the cases discussed in Section 4.5.2, where the signal space was one-dimensional, the shift could only be along a single axis. Fig. 4.45 shows the analogous situation for QPSK. It can be shown that for

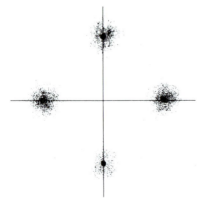

Fig. 4.45. Effect of noise in two-dimensional signal space.

additive, white, Gaussian noise with zero mean, a shift is equally likely in any direction, and the probability of a given shift also follows a Gaussian distribution. This is indicated in the figure by the circular symmetry of the shading, and the way it becomes lighter as the distance from the ideal point increases. In a three-dimensional signalling space, points could be displaced due to noise in three dimensions, and the distribution would have spherical symmetry. Even more importantly, however, this 'smearing' of points in signal space can be quantified. Again, a detailed proof is inappropriate here, but the same result applies as in the one-dimensional case: the standard deviation of the displacement of points in *signal space* around the ideal location is simply $\sqrt{(\eta/2)}$, where η is the (single-sided) power spectral density of the AWGN.

In the one-dimensional case, an error occurs if a detected pulse is allocated by the receiver to a point in signal space nearer to another possible state than to the one corresponding to the symbol actually transmitted. The same principle applies to higher-dimension signal space: in fact, the result derived in Section 4.5.1 holds for any binary system, providing matched filter detection is used. That is, the probability of error is given by

$$P_{\mathrm{e}} = Q[\sqrt{(E_1/2\eta)}]$$

where $\sqrt{E_1}$ is the *separation* in signal space between the points representing the two symbols used and, as before, η is the (single-sided) power spectral density of the AWGN. This (theoretical) formula holds precisely for any binary signalling scheme using optimal detection, and in general the separation of states in signal space is a direct measure of the error probability.

The signal space model gives a clear illustration of a number of important trade-offs in telecommunications already mentioned. For example, as with increasing the number of voltage levels in baseband signalling, increasing the number of states in a given signal space permits an increased rate of information transmission for a given signalling rate and hence bandwidth.

This could mean, for example, increasing the number of amplitudes in ASK, or phase states in PSK, or both in a more general QAM scheme.

But to maintain the same spacing in signal space with an increased number of symbols, the average pulse energy will need to be increased, thus requiring increased power for a given error rate. Going from binary to higher order FSK avoids the necessity for increasing the pulse energy and hence the transmitter power, as shown in Fig. 4.46, where three states in three-dimensional signal space retain the same spacing and the same energy per pulse as two states in two dimensions. The disadvantage here, though, is that increased bandwidth will be required to accommodate a new pulse using a new frequency, illustrating yet again the classic S/N

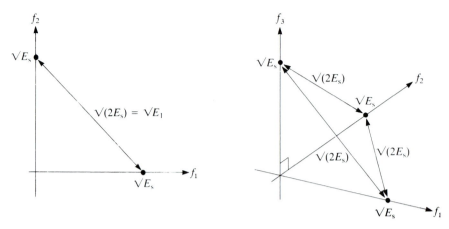

Fig. 4.46. For a given symbol energy E_s, the distance between any two states in ternary FSK is the same ($\sqrt{E_s}$) as between the states in binary FSK.

ratio v. bandwidth trade off. There are few hard and fast rules, therefore, about how to select an appropriate scheme for digital transmission. In any practical situation, the final choice will take into account receiver (and to a lesser extent transmitter) complexity, permissible error rate, channel frequency response and noise characteristics, transmitter power limitations, cost, available bandwidth, and many other factors.

4.6.4   A comparison of binary sinusoidal signalling schemes

Fig. 4.47 gives a comparison of results for binary OOK, PSK and FSK, assuming a signalling rate of B states s^{-1}. The same separation $\sqrt{E_1}$ is assumed between signalling states, so the same error rate $P_e = Q[\sqrt{(E_1/2\eta)}]$ is to be expected in each case. The energies of the various individual pulses will differ, however, being equal to the square of the distance from the origin of the appropriate point in signal space. The mean and peak powers at the receiver therefore vary according to the signalling scheme adopted, and are included on the diagram.

As an alternative way of comparing schemes, the error probability can be re-expressed in terms of E_s, the energy per symbol. That is, we assume that the various schemes use pulses of the same energy, and examine the consequences for error rate. For OOK the energy per pulse is equal to E_1, so we have:

$$P_e = Q[\sqrt{(E_s/2\eta)}]$$

For PSK,

$$\sqrt{E_s} = \sqrt{E_1}/2$$

so:

$$P_e = Q[\sqrt{(2E_s/\eta)}]$$

Finally, for FSK

$$\sqrt{E_s} = \sqrt{(E_1/2)}$$

so:

$$P_e = Q[\sqrt{(E_s/\eta)}]$$

Some conclusions from the preceding results are:

1 For a given error rate, PSK requires the lowest mean receiver power. Conversely, for a given mean power, PSK gives the best error rate.

2 PSK requires only half the mean power of OOK for a given error rate. (This is the general relationship between any antipodal scheme and its unipolar counterpart.)

3 For a given error rate, OOK requires the same mean power as FSK but twice the peak power (during data 1s).

4 FSK requires a larger bandwidth than OOK or PSK.

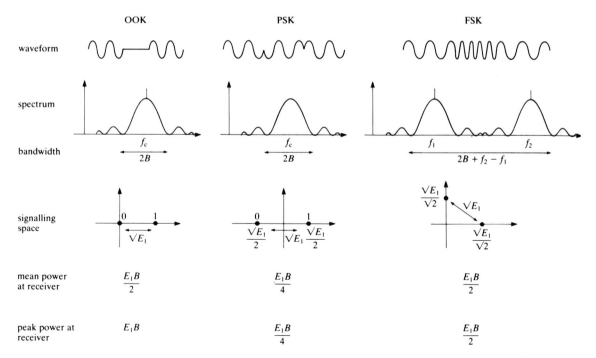

Fig. 4.47. A comparison of theoretical results for binary OOK, PSK and FSK.

It would seem that PSK is better than OOK or FSK on all counts. However, bear in mind that the above results are theoretical optimum results, based on optimal (matched filter) detection. Which scheme is the best solution for any given application will also depend on the practicalities of transmitter and receiver design. There is little difference in transmitter complexity between any of the three basic techniques, so receiver considerations tend to dominate. In general, too, there will be a number of options for receiver design for any particular modulation scheme, and it is again often possible to trade off complexity against receiver error performance. Receiver design cannot be considered here in any depth, but a useful insight into some of the important issues can be gained from a brief look at common techniques.

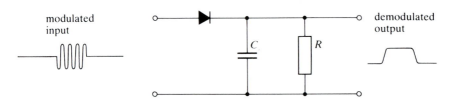

Fig. 4.48. Envelope detector for amplitude-modulated signals.

One of the simplest approaches to detecting sinusoidal pulses, and one which can be applied to OOK, ASK and FSK (but not PSK) is *envelope demodulation*. The sinusoidal pulse is rectified and smoothed as shown in Fig. 4.48. For an OOK or ASK waveform this process can be followed by threshold detection to make a decision about the particular symbol transmitted. For FSK, recall that such a waveform can be considered as interleaved OOK. Two OOK waveforms can be extracted by appropriate filtering, and envelope demodulation followed by threshold detection can then take place as before (Fig. 4.49).

An FSK scheme using comparatively widely spaced frequencies, as was illustrated in Fig. 4.37, is being assumed here.

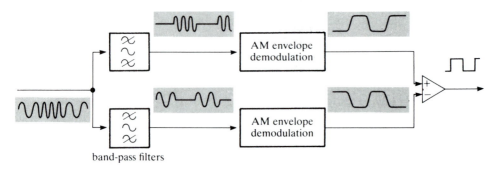

band-pass filters

Fig. 4.49. Dual-filter envelope detection for binary FSK.

Any process involving envelope detection expressly ignores the phase information of the sinusoidal waveform, a fact which has two important consequences. The first is that it cannot be applied to any scheme in which the digital symbols are distinguished by means of phase relationships (PSK, QAM, for example). The second (which is not so obvious, and really needs to be brought out through appropriate mathematical analysis), is that the loss of such phase information results in a sub-optimum signal to noise ratio, and hence error rate, at the threshold detector.

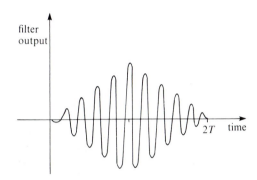

Fig. 4.50. Matched filter response to a single sinusoidal pulse.

The matched filter approach to signal detection takes into account such phase relationships – recall that it is a crucial feature of the matching process that all incoming signal components reinforce in phase at the instant of sampling. Fig. 4.50 shows the response of a matched filter to an input sinusoidal pulse. As expected, the output reaches a peak (equal to the pulse energy) at time $t = T$. Note, however, the potentially enormous consequences of a small timing error in sampling the filter output, even supposing that the ideal matched filter could be realised. There are various ways of avoiding this problem. In OOK, ASK or FSK the matched filter(s) could be followed by an envelope detector to eliminate the oscillations yet still detect the peak; this would have the drawback (as before) of giving a less than optimal signal to noise ratio, and it still cannot be applied to PSK since the vital phase information remains lost. An alternative approach, with a number of practical applications, is to use a correlation detector like that introduced in Section 4.5 and shown again in Fig. 4.51. This has various advantages: the output of the integrator does not oscillate in the same way as that of the corresponding matched filter, so the timing of the sampling instant is not so critical; and the signal to noise ratio retains its optimal value (since at the sampling instant the

Try sketching the integrator output for a sinusoidal pulse $p(t)$**. Compare your sketch with Fig. 4.50.**

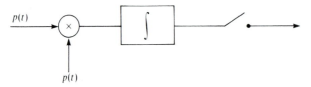

Fig. 4.51. The principle of a correlation detector.

correlation detector gives an output S/N ratio equal to that of a matched filter).

However, all such *coherent* or *synchronous* techniques remain sensitive to timing errors between the received pulse and the locally generated copy – hence their design complexity in comparison with *non-coherent* techniques such as envelope demodulation. Coherent detection methods rely on extracting the carrier, in the correct phase relationship, from the incoming signal. They can be applied directly to some sinusoidal signalling techniques – but not to PSK, since there is no such carrier component present in a PSK spectrum. While it is still possible to derive a sinusoid at the carrier frequency by means of non-linear processing, a phase uncertainty remains: in particular, uncertainty as to whether the derived carrier sinusoid is in-phase or anti-phase. This latter problem can be overcome by using a differential coding technique and the topic is taken up again in Chapter 6.

4.7 Summary

This chapter has applied the modelling tools of Part 1 to the problem of how pulses can be transmitted so as to minimise error. A theoretical maximum signalling rate (Nyquist rate) of $2B$ baud is possible over a bandlimited channel of bandwidth B Hz. For a baseband system this implies received pulses at the threshold detector approximating to a $(\sin x)/x$ waveform. Such pulses are susceptible to intersymbol interference (ISI) owing to their highly oscillatory waveform. A common practical solution is to tailor the received pulse shape (at the expense of additional bandwidth) so that its spectrum approximates that of a raised-cosine: if 100% raised-cosine shaping is used then the maximum signalling rate in bauds is equal to the channel bandwidth in hertz.

An alternative approach to ISI is to allow it to occur in a controlled way, by means of partial response signalling. Two duobinary schemes were briefly presented which allow signalling at the Nyquist rate – but only at the expense of greater susceptibility to noise.

Receivers often incorporate digital equalisers to 'mop up' slight intersymbol interference after initial filtering at the receiver. A $2n + 1$ term transversal digital filter can force n zeros either side of a central peak. Such equalisers are often made adaptive by the use of microprocessor control.

A powerful new modelling tool was introduced, which relies on the extension of the concept of orthogonality (introduced in Chapter 3) to finite-duration pulses. Two signals $s_1(t)$ and $s_2(t)$ are said to be orthogonal in the interval t_1 to t_2 if

$$\int_{t_1}^{t_2} s_1(t)s_2(t)\,dt = 0$$

Linear combinations of orthogonal signals can be represented as points in signal space: the distance between signal space points representing digital symbols is a direct measure of the susceptibility of the symbols to misinterpretation, and thus of system error rate.

A useful ideal against which to measure the performance of any system is provided by the optimal detector which, for a given pulse, maximises the signal to noise ratio at the instant of threshold detection. Optimal detectors are: (i) the correlation detector, in which the incoming pulse is multiplied by a replica and then integrated over the symbol period before threshold detection; and (ii) the matched filter whose impulse response $h(t)$ is matched to the pulse shape $p(t)$ by the expression $h(t) = p(T-t)$, where T is the signalling interval. In frequency-domain terms, $|H(\omega)| = |P(\omega)|$ and $\text{Arg}\,H(\omega) = -\omega T - \text{Arg}\,P(\omega)$. For AWGN with zero mean at the input, the noise at the optimal detector output is Gaussian with zero mean with a mean-square value numerically equal to $E\eta/2$, where E is the energy of the pulse and η is the (single-sided) power density of the input noise. The overall error probability for baseband binary signalling is then given by $P_e = Q[\sqrt{(E/2\eta)}]$. This result applies both to on–off signalling (using a pulse with energy E to represent one symbol, and no pulse to represent the other), and to antipodal signalling using pulses with energy $\pm E/4$ (in both cases the signal states are separated by $\sqrt{E}$ in signal space).

These ideas were introduced in the context of baseband signalling, but apply equally to systems using sinusoidal pulses (modulated schemes) for on–off keying (OOK), amplitude shift keying (ASK) phase shift keying (PSK) or frequency shift keying (FSK). Again, pulses can be represented in an appropriate signal space, and practical schemes compared against the theoretical ideal of a matched filter or correlation detector. For optimal detection using pulses with energy E_s, the following theoretical results apply for overall error probability:

For OOK

$$P_e = Q[\sqrt{(E_s/2\eta)}]$$

For PSK

$$P_e = Q[\sqrt{(2E_s/\eta)}]$$

For FSK

$$P_e = Q[\sqrt{(E_s/\eta)}]$$

The choice of scheme for any practical application will also be influenced by receiver (and to a lesser extent transmitter) complexity and cost.

5

Line codes

5.1 Introduction

In the layered model introduced earlier line coding is immediately above pulse generation and below channel coding. The input from the channel coding layer is an arbitrary binary sequence plus a timing waveform, and the output to the pulse generation layer can be thought of as a (structured) sequence of impulses.

In a real system line coding is generally very closely connected with pulse generation and it would often not be possible to separate completely the electronic circuits performing line coding from those generating the pulses. The close connection between line codes and pulse generation is reflected in the presentation in this chapter, where line coded *waveforms* will be illustrated (assuming, generally, rectangular baseband pulses), rather than the sequences of impulses which form the conceptual output of the line code layer.

Although there is a theoretical background to the analysis of line codes (some of which is discussed here), in practice the evolution of line codes has involved a large element of pragmatism. New line codes have often been designed for use in specific systems, with features to combat particular problems in those systems.

Most line codes have three major functions. First, they allow a baseband signal to be conveyed over a channel with d.c. blocking. This is achieved by ensuring that the coded data does not build up short-term d.c. offsets. In the frequency-domain it is necessary for the spectrum of the coded data to *fall-off* to zero at d.c.: it is *not* sufficient to ensure that the coded data has no d.c. component. For example, bipolar binary NRZ (that is, a binary waveform where a 1 is represented by a positive voltage and a 0 by an equal negative voltage) has no d.c. component because, assuming the data has on average the same number of 1s and 0s, it has equal positive and negative voltage. Nevertheless, data encoded by bipolar NRZ can cause problems when conveyed over a channel with zero d.c. response. Any long runs of 0s or 1s will result in 'droop' or *'baseline wander'* as nominally constant voltage levels drift up or down due to capacitors charging or discharging (Fig. 5.1).

The second major function of most line codes is to ensure that the coded

Even so-called 'baseband channels' are normally a.c. coupled – by either capacitors or a transformer. This is because the earth voltage (nominal 0 V) can vary substantially between distant locations (or even between floors in a single building). D.C. blocking is required to prevent 'earth currents' from flowing.

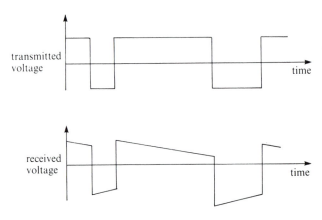

Fig. 5.1. Baseline wander due to a.c. coupling.

data contains enough timing information for a clock waveform to be extracted at the receiver. Timing information can be recovered from data provided that transitions in the signal waveform occur sufficiently frequently. By ensuring that there are never long periods without a transition in voltage levels a line code effectively incorporates timing information into the bit stream itself.

Finally, most line codes allow error detection to be performed. This is possible because the coding introduces redundancy into the data. Certain data sequences are allowed, others are not. A single error will (in most codes) turn an allowed coded data sequence into one which is forbidden, and can thus be detected (if not corrected).

5.2   Code descriptions

5.2.1   Uncoded data: unipolar and bipolar

In the layered model the distinction between unipolar and bipolar is a function of the line code layer: for bipolar the output to the pulse generator is a sequence of positive and negative impulses, for the unipolar the output is an impulse for a binary 1, no impulse for a 0.

However, a system which uses either unipolar or bipolar signalling with no further line coding is often said to use uncoded data.

5.2.2   Alternate mark inversion

An example of a simple line code which attempts to provide the functions outlined in Section 5.1 is Alternate Mark Inversion (AMI).

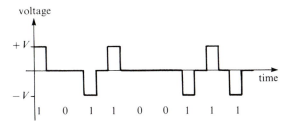

Fig. 5.2. An AMI waveform.

In AMI, zero volts is used to represent binary 0, and successive binary 1s are represented by alternate, equal positive and negative voltages, usually as half width (RZ) pulses. A typical AMI waveform is illustrated in Fig. 5.2. AMI prevents droop since the longest time spent at any non-zero voltage level is equal to the width of a single pulse. The timing content of the data is also improved since strings of binary 1s are encoded as alternating pulses – and there is therefore a transition between each pulse. AMI alone does not guarantee that the encoded data has sufficient timing content, however, because long strings of binary 0s would be encoded by constant zero volts. Some systems nevertheless use AMI if the input data is known to have a guaranteed limit to the length of runs of binary 0s. This is the system used on some PCM links in the USA. Alternatively, precoding can be used to eliminate, or at least reduce the probability of, long strings of 0s (scrambling – see Chapter 6 – is sometimes used for this purpose). More complex codes, some of them modifications of AMI, can guarantee a good timing content.

5.2.3   Codes based upon AMI

HDB3 This code is a modified form of AMI which gives a guaranteed timing content by ensuring that no more than 3 symbol periods can elapse without a voltage transition.

As explained above, the problem with the timing content of AMI occurs only during strings of data 0s. If the input data contains no string of 0s longer than three, HDB3 is identical to AMI. If, however, a string of four 0s is to be transmitted, in HDB3 the last of the 0s is represented by a pulse, but with a polarity which violates the AMI coding rule: that is, it has the *same* polarity as the previous pulse instead of the opposite polarity (Fig. 5.3). This added pulse is known as a 'V' (violation) pulse. The receiver still recovers the intended data because it 'knows' that a violation pulse always represents a data 0.

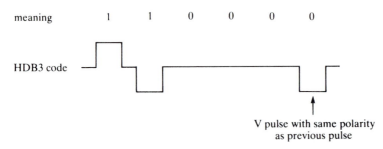

Fig. 5.3. Violation pulses in HDB3 coding. (Note that the convention used in this book is to draw bit streams and waveforms with the first bit to be transmitted on the left.)

The use of violation pulses alone as just described would be unsatisfactory however, because although it certainly would ensure the required timing content, it would do so at the expense of re-introducing the problems of baseline wander. This is because successive violation pulses could have the same polarity, as in the encoding shown in Fig. 5.4, which leads to a build-up of d.c. offset.

Such a build-up is prevented in HDB3 by a further modification to the coding rule. Another additional pulse, this time in *agreement* with the AMI encoding rule, is inserted in place of the first of a group of four 0s whenever necessary to prevent two successive violation pulses having the same polarity. The extra pulse is referred to as a 'B' (balancing) pulse. The correct HDB3 encoding of the data of Fig. 5.4 is therefore as in Fig. 5.5 (assuming that a B pulse was not required in the first string of four 0s). Notice that an encoder will not know whether a 0 is to be encoded by a B pulse or by zero volts until it has received the next three input bits (and checked whether they are all 0s). So each input bit has to be stored in the encoder until three subsequent bits are received.

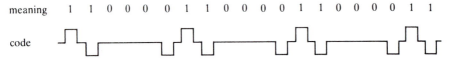

Fig. 5.4. The build-up of a d.c. offset as a result of successive violation pulses of the same polarity.

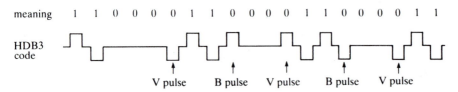

Fig. 5.5. Balancing pulses in HDB3. The balancing B pulse ensures that successive V pulses have the opposite polarity. Notice that a pulse representing a 1 always has the opposite polarity to the previous pulse – even if the previous pulse was a V pulse.

A receiver can recognise a B pulse by the fact that it is always followed by a violation pulse with only two intervening 0s.

CMI (Coded Mark Inversion) CMI (Fig. 5.6) is a binary code which can be thought of as another modification of AMI. It is used at high signalling rates in preference to HDB3 because it uses simpler circuitry for the encoders and decoders – a feature which is more significant at high signalling rates because high frequency circuits tend to be complex and expensive. The simplicity is partly because CMI is a binary rather than a ternary code and partly because the coding rules are simpler.

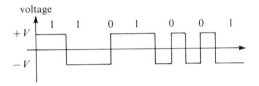

Fig. 5.6. A CMI waveform.

1s are represented by alternating pulses as for AMI, using full-width pulses. 0s are represented by a half period at the negative pulse voltage followed by a half period at the positive pulse voltage.

EXERCISE 5.1

What would be the AMI, HDB3 and CMI encodings of the binary sequence 1000111000000001100001?

5.2.4 Code parameters: Radix, redundancy and efficiency

Code radix The radix of a code is the number of different signalling states that it uses. In the layered model, it is the number of different strengths of impulse generated by the code (including zero strength if appropriate). For baseband systems the radix becomes the number of voltage levels used. For example, both unipolar and bipolar use just two voltage levels so have radix 2. AMI uses three voltage levels (positive, zero and negative) so has radix 3. Most baseband line codes have either radix 2 or 3. The receiver design becomes much more complex for higher radix. In particular, the decision thresholds at the receiver need to be varied dependent upon the line length: this is not necessary for radix 2 or 3 codes, as shown in Fig. 5.7.

In practice the thresholds *may* be varied in a system using a radix 3 code, to optimise the error rate.

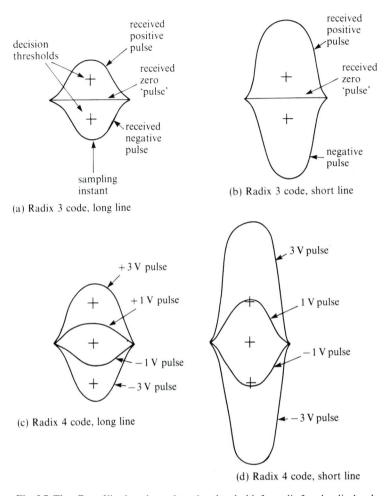

Fig. 5.7. The effect of line lengths on detection thresholds for radix 3 and radix 4 codes.

Fig. 5.7(a) shows the received pulses of a radix 3 code, and the corresponding decision thresholds. Fig. 5.7(b) shows the received pulses of the same system for a shorter line, using the same decision thresholds. The decision thresholds are not at the ideal voltage for this line length (so they would not give the best noise immunity) but the noise free pulses would still be correctly detected. Figs. 5.7(c) and 5.7(d) make the same comparison for a radix 4 code, using the transmitted pulses $+3, +1, -1$ and -3 volts (Table 5.7). In this case the decision thresholds used for the longer line simply would not work for the shorter line, even in the absence of noise. The problem is that even the 1 V pulses exceed the threshold expected at the receiver for the transmitted 3 V pulses.

Redundancy and efficiency In Chapter 4 the expression $\log_2 M$ was introduced as the factor relating the data rate in bits s^{-1} to the signalling

To utilise this capacity to the full, special non-redundant ways of allocating the message information to the three available states would have to be designed. Codes such as AMI, which are ternary but only carry the same amount of information as binary codes, are sometimes called *pseudoternary* codes. For a full treatment of how to quantify information as 'bit's worth', the reader is referred to the discussion of information theory in Schwartz.

rate in bauds for an M-ary code, where M was assumed there to be a power of 2. In fact, the expression $\log_2 M$ has a wider significance, and can be thought of as a general measure of the maximum information it is possible to transmit per symbol in a given code, on average. Thus a binary code can carry at most one bit's worth ($\log_2 2$) of information per symbol on average; a quaternary code, two bit's worth ($\log_2 4$); an 8-ary code, three bit's worth ($\log_2 8$); and so on. The higher the radix of a code, the more information per symbol it can convey. In fact, the idea can be extended to codes using numbers of states which are *not* powers of 2. So, for example, a ternary (3-level code) such as AMI can in theory transmit $\log_2 3 = 1.58$ bit's worth of information per symbol. In practice, it is not always possible or desirable to use the full information-carrying capacity of a code. Such codes are said to be redundant, an important feature which can allow errors to be detected or even corrected, as will be discussed below. AMI is a simple example, where each binary input bit is coded as a single ternary symbol; it therefore carries only 1 bit's worth of information per symbol instead of the theoretical maximum of 1.58.

The *redundancy* of a code is quantifiable as:

$$redundancy = \frac{information\ per\ symbol\ available - information\ per\ symbol\ used}{information\ per\ symbol\ available}$$

This is the definition of redundancy used by the CCITT; it differs from that used by some texts on line codes (such as Bylanski and Ingram, Waters). The alternative definition has the information per symbol *used*, rather than information per symbol *available*, in the denominator.

AMI, for example, has redundancy $(1.58 - 1)/1.58 = 0.367$.

The efficiency of a code is just an alternative measure of the same quantity. Efficiency is defined as:

$$efficiency = \frac{information\ per\ symbol\ used}{information\ per\ symbol\ available}$$

Efficiency and redundancy are therefore related by

$$Efficiency = 1 - redundancy$$

Chapter 4 explained how at higher signalling rates the received pulses need to have a higher amplitude in order to maintain the pulse energy. To achieve the higher amplitude the cable needs to be shorter. (See the example in Section 4.5.1 on page 124).

The efficiency of AMI is 0.633 or 63.3%.

The significance of the efficiency and redundancy of a code is that, for a given code radix and signalling rate, a more efficient (less redundant) code can convey a greater information rate. Or, equivalently, if a link is required to convey a given information rate, then for a given code radix a more efficient code will require a lower signalling rate. There is an advantage in keeping the signalling rate as low as possible because it allows greater cable lengths between repeaters. Notice, however, that the redundancy and efficiency of a code refer only to its information-carrying capacity; they say nothing about other code characteristics which may be important in any given application. For example, the reduced efficiency of AMI compared to NRZ is accepted in many applications because of the immunity of AMI to baseline wander.

5.2.5 Block codes

Block codes, like codes based on AMI, are designed to prevent baseline wander and to have a good timing content. They do so by using a different principle from AMI and require more complex encoding and decoding circuits. In exchange for this extra complexity, they can be designed to provide a higher efficiency. As discussed in the previous section, the advantage of higher efficiency is that by reducing the signalling rate it allows longer cable lengths between repeaters. So the application of block codes tends to be in long transmission systems, where the extra complexity in the encoding and decoding circuits is justified by the saving in the number of repeaters required.

The basic principle behind block codes is that blocks of input bits are encoded to give blocks of code symbols, rather than a single input bit being encoded to one code symbol as with all the codes discussed so far.

The encoder for a block code (Fig. 5.8) has first to break the incoming data bit stream into a block of bits (sometimes called words) and then

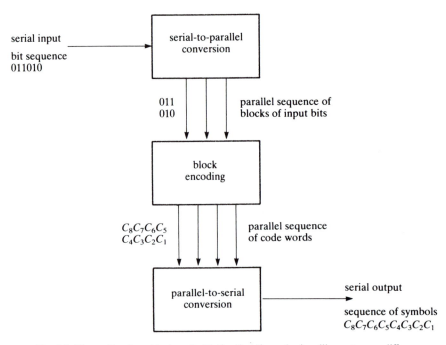

Fig. 5.8. The coding for a block code. Notice that the code signalling rate may differ from the input data rate.

Table 5.1. *A 3B4B coding table*

	Output			
Input	Negative	0	Positive	Disparity
001		$--++$		0	
010		$-+-+$		0	
100		$+--+$		0	balanced
011		$-++-$		0	words
101		$+-+-$		0	
110		$++--$		0	
000	$--+-$		$++-+$	± 2	unbalanced
111	$-+--$		$+-++$	± 2	words

encode each block of input bits to a block of code symbols (the code words). Finally it transmits the code words serially. A ternary block code is often described as an nBmT code, where n binary bits are encoded to m ternary symbols. Similarly, a binary block code can be denoted as an nBmB code.

Note that the incoming data need have no inherent block structure: the position of the initial block boundary chosen by the encoder is arbitrary and subsequently the encoder takes successive sequences of n adjacent bits.

EXERCISE 5.2

What are the radix, redundancy and efficiency of:
(a) CMI
(b) 3B4B
(c) 4B3T?

3B4B Block codes are described by means of a *coding table* which shows how each possible input word is encoded. Table 5.1 is a coding table for a 3B4B code. '+'s and '−' have been used for positive and negative pulses respectively. It is more usual to see binary code tables written out as 1s and 0s, but for the explanation of the principles of block codes it is convenient to assume that 1s are being represented by a positive pulse and 0s by a negative pulse.

The left hand column of Table 5.1 lists all possible 3-bit binary inputs; the next three columns contain the possible binary outputs according to the coding rule. Some inputs have a choice between two codings; the reason for this will become clear.

The fifth column of Table 5.1 contains the code word disparity. The disparity, or digital sum, of a code word is a count of the imbalance between the number of positive and negative pulses, and is therefore a normalised measure of the mean voltage of the word. A word with equal numbers of positive and negative pulses would have a zero disparity and zero mean voltage, and is described as balanced. Unbalanced words can have either positive disparity (more positive pulses than negative and therefore a positive mean voltage) or negative disparity (more negative pulses than positive, and therefore a negative mean voltage).

To ensure that there is no short-term mean voltage offset, balanced code words are used in Table 5.1 for the output where possible; that is, for input words 001 to 110. The remaining input words have to be encoded to unbalanced words. To maintain the zero offset each of these input words can be encoded to either a word with a positive mean voltage or a negative mean voltage. The choice between the two for a particular input word is made at the time of encoding, in such a way to reduce the mean voltage offset.

Fig. 5.9 illustrates how this works. The encoder keeps a record of the running sum of disparity (the running digital sum) of each transmitted word. That is to say, the coder has a register which records a number, called the running digital sum, to which it adds the disparity of each word as it is transmitted. The encoder consults this sum whenever an unbalanced word must be transmitted. If the sum is negative, the positive disparity option is chosen. Whenever it is zero or positive, the choice is for the negative disparity option. Clearly, if the input word is encoded by a

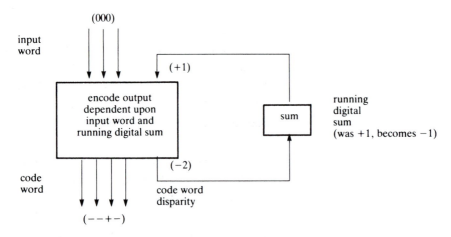

Fig. 5.9. Use of the running digital sum in the encoding of a block code. (Symbols in brackets show an example encoding).

Table 5.2. *Encoding of data by a 3B4B code*

Input word	Running digital sum before	Coded to	Running digital sum after
111	0	− + − −	−2
011	−2	− + + −	−2
000	−2	+ + − +	0
000	0	− − + −	−2
010	−2	− + − +	−2
100	−2	+ − − +	−2

balanced word, then no choice needs to be made and the running sum is unchanged.

Following this coding rule and assuming that the register initially contains $+1$, the running digital sum will always be either $+1$ or -1 between the transmission of words. So the encoder can be regarded as having two different states, defined by the current running digital sum.

The absolute values of the disparity states are arbitrary. If the starting state is defined to be, say, 0, then the encoding rule would lead to the running digital sum being always 0 or 2. The important thing is the number of states – in this case two – and the idea that the coding rule provides a 'negative feedback' which keeps the mean voltage in check.

EXAMPLE

Assuming that the encoder starts with the running digital sum equal to zero, what is the following data encoded to?

111 011 000 000 010 100

Solution. The encoding is shown in Table 5.2, and the output sequence is therefore

− + − − − + + − + + − + − − + − − + − + + − − +

Decoding is performed by first breaking the incoming serial data into words of four symbols. (The decoder needs to 'know' where to break the serial data in order to get the correct blocks of code symbols: this is discussed in Section 5.4) Then, by 'reading backwards' from the coding table, the decoder can deduce what the input binary word was. A practical transmission system would perform this by means of a look-up table consisting of all the possible received code words with their corresponding

Table 5.3. *Decoding table for a 3B4B code*

Received code word	Decoded to
− − − − *	001
− − − + *	000
− − + −	000
− − + +	001
− + − −	111
− + − +	010
− + + −	011
− + + + *	011
+ − − − *	100
+ − − +	100
+ − + −	101
+ − + +	111
+ + − −	110
+ + − +	000
+ + + − *	111
+ + + + *	110

*These code words are 'forbidden' words.

decoded binary forms. A decoding table for this 3B4B code is given in Table 5.3.

The code words marked by an asterisk in Table 5.3 are 'forbidden words', i.e. they are words which are not used in the coding table. However, they might appear in the coded data arriving at a decoder because they might have been created by errors in the data stream. The decoder therefore needs to know how to decode them, so an entry for them is included in the decoding table. The choice of decoding for these words is discussed in Section 5.3.3.

EXERCISE 5.3

How would the following data be encoded using the 3B4B code of Table 5.1?:

100001100001000001011100000001

1B2B codes Although CMI was described above as a modification of AMI, it can also be interpreted as a 1B2B code, with the code table of

Table 5.4. *CMI coding table*

Input word	Code word			Disparity
	Positive	0	Negative	
1	+ +		− −	±2
0		− +		0

Table 5.5. *Manchester code coding table*

Input	Code word	Disparity
1	− +	0
0	+ −	0

Table 5.6. *Manchester bi-phase mark*

Input	Code words	
	Positive disparity	Negative disparity
0	+ +	− −
1	+ −*	− +*

*These code words are balanced, but are treated as unbalanced insofar as + − is only transmitted if the running digital sum is negative, and − + if the running digital sum is positive.

This feature of immunity to complementing can also be achieved by differential coding (see chapter 6). In fact another way of generating Manchester bi-phase mark is by coding 1 to 01, and 0 to 11, then differentially coding the resultant code sequence.

Table 5.4. Other 1B2B codes are possible, most notably Manchester code which has the code table of Table 5.5. Although CMI and Manchester code are very similar, there are differences in their properties which make them appropriate for particular applications. For example, block alignment is very easy with CMI whereas Manchester code has the greater immunity to baseline wander.

One other 1B2B code of interest is that which is sometimes called Manchester bi-phase mark. This has the code table of Table 5.6, and has the particular feature that complementing the coded data (interchanging + and −) does not affect the output at a decoder.

EXERCISE 5.4

How would the following data sequence be encoded in CMI, Manchester and Manchester bi-phase mark codes? 11011011001101

5.2.6   High efficiency codes

There are circumstances where it is more important to have a code with a high efficiency than to have a guaranteed good timing content or to protect against baseline wander. Examples of codes used in such circumstances are 24B1P and 2B1Q.

24B1P The 'P' in this code stands for parity: the coding consists of adding one extra binary symbol (the 'parity bit') after 24 input bits. (So an alternative description could be as a 24B25B block code.) The parity bit is a 0 if the 24 bits in the input have even parity (i.e. if there is an even number of 1s among them) and 1 if they have odd parity (if there is an odd number of 1s among them). In other words the 25 symbol block always has even parity.

24B1P neither controls baseline wander nor ensures a good timing content: all it does is allow error detection. It is, however, very efficient (96%). It is used in some optical fibre transmission systems, for example in the transatlantic optical fibre link known as TAT-8.

2B1Q 2B1Q is just non-redundant quaternary. Pairs of input bits are encoded to one of four states. Using rectangular pulses these would just be four voltage levels, such as in Table 5.7.

Again, 2B1Q neither controls baseline wander nor ensures a good timing content. It does not even allow error detection. Furthermore, being a radix 4 code, it requires complex line length dependent equalisation at a receiver for the reasons discussed in Section 5.2.4.

Despite all these disadvantages there are situations where it is used – because of the fact that it halves the signalling rate by coding two input bits to a single channel symbol. In particular, it is one of the codes which may be used to provide a digital connection between customer terminals and a local exchange over wires installed originally for a single analogue speech channel. The data rate required over such a link to comply with CCITT recommendations for ISDN basic rate access is, including framing overheads, 160 kbits s^{-1}. It has been shown that the length which can be achieved in this application is limited by near-end crosstalk (NEXT) (Adams and Cook, Lechleider). Because crosstalk is more

Table 5.7

00	$-3V$ volts
01	$-V$ volts
10	$+V$ volts
11	$+3V$ volts

serious at higher frequencies, longer links are possible when 2B1Q is used than when any binary or ternary code is used. The high complexity needed at the receiver to make use of 2B1Q is, paradoxically, less serious in this application because of the large numbers of customer to local exchange links which would be involved. The complex circuitry can be designed into a dedicated integrated circuit, and the development cost amortised over the large numbers produced. The production cost of the dedicated IC is then quite small – and to some extent independent of the complexity of the circuit on the device.

5.3 Decoding and error detection

Apart from more-or-less random errors caused by noise, interference and crosstalk, there will be errors caused by system failures, component failures, cut or damaged wires etc. When these occur, it is important for the communication link to detect them so that the fault can be rectified. For this reason, error detection is a vital part of a transmission link. Furthermore, for a transmission link which includes repeaters, error detection at each repeater simplifies fault finding by locating the fault to within one repeater section.

When errors have been detected, it is important for the decoder to minimise their effect: that is to say, correct them if possible and, if not, attempt to ensure that they do not have a 'knock-on' effect and cause further errors.

5.3.1   Error detection at a decoder

Most of the codes discussed in the previous section have some degree of redundancy. It is possible to make use of this redundancy to check for errors. All the codes use some sort of coding rule, and the receiver can check to see if the data is consistent with the rule. Violations of the coding rule indicate that errors have occurred along the route.

Take the simple example of AMI. The data 110011 could be encoded in AMI as:

$$+ - 00 + -$$

Suppose an error occurs which results in the receiver interpreting the third symbol as a positive pulse, so that it receives

$$+ - + 0 + -$$

which it would decode to 111011. But this means that two successive 1s have apparently been encoded by positive pulses – violating the AMI coding rule (which requires 1s to be represented by alternating polarity pulses). So the receiver can flag the presence of an error. It is not possible for the receiver to correct the error because there is more than one data sequence which would create the same received sequence when corrupted by a single error. For example, 111111 could be encoded as

$$+ - + - + -$$

and a single error changing the fourth symbol from a $-$ to a 0 would generate $+ - + 0 + -$, the same as above.

If you consider in turn the various ways in which a single error can occur (0 changes to a positive or negative pulse, positive pulse changes to 0 or a negative pulse, negative pulse changes to a 0 or positive pulse) you will see that they will all eventually cause a violation. Any single error is therefore detected eventually, but there can be a delay. For example, if an error occurs in the third symbol of

$$+ - 00000000000 +$$

turning the first 0 into a positive pulse, it will not be until the end of the string of 0s that the error can be detected.

Multiple errors will not necessarily be detected. In the example just given, if a second error occurs before the end of the string of 0s (say at the tenth symbol) and if that error turns the 0 into a negative pulse, then the resulting data will become

$$+ - + 000000 - 0000 +$$

This will be decoded to give the wrong sequence, but the receiver will not detect an error. So when errors occur randomly, at low error rates virtually every error will be detected at the receiver, but at higher error rates some errors will 'cancel out' before the receiver has detected them.

Typically this results in error detection performance with the characteristics of Fig. 5.10 which shows a plot of the measured error rate against the actual error rate, assuming random errors and random data. At high error

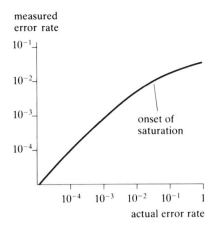

Fig. 5.10. Typical error detection system performance. ('Error rate' is the proportion of symbols in error; so an error rate of 10^{-2} implies that 1 symbol in every 100 is in error.)

Some error detection schemes (such as Even Mark Parity, Section 5.3.2) can, however, actually result in the measured error rate falling for very high error rates. This can be a problem if the apparent error rate falls below the fail threshold.

rates the error detector starts to saturate as the errors are close enough to interfere with each other.

In a practical transmission system, the link is said to have failed when the error rate is found to be greater than some predefined threshold (generally in the range 1 in 10^6 to 1 in 10^3, depending on the particular application or where in the hierarchy the link is contained). It is not necessary to measure the error rate above this threshold, so saturation of the error detector is not important – provided it does not begin until an error rate greater than the fail threshold of the link.

At first sight it might seem that because HDB3 deliberately uses 'violation pulses' to incorporate timing information into the waveform, error detection by looking for code violations would not be possible in HDB3. In fact, this is not so. HDB3 still uses a fully defined coding rule, and any single error can be detected just as for AMI. (If this seems improbable, try it out by putting in single errors to strings of HDB3 encoded data).

The technique of detecting violations of a coding rule also applies to block codes. Taking the 3B4B code of Section 5.2.5 as an example, consider the encoding of 111001010000000. This becomes (assuming an initial running digital sum of -1)

$$+-++\ --++\ -+-+\ --+-\ ++-+$$

Suppose the sixth symbol is incorrectly received as $+$, then the second received word would be $-+++$. This is a forbidden word, that is to say it is a 4-symbol word which is not used by the code, so the decoder knows an error must have occurred. Not all errors create forbidden words, however. For example suppose an error changes the fifth symbol to $+$.

This makes the second word $+-++$ which is still a legitimate code word. However, the received sequence is now

$$+-++ \quad +-++\ldots$$

where the first two words both have positive disparity, something which should never happen according to the encoding rule used by this 3B4B code. The sequence $+-++$ is an encoding of 111, but when the running digital sum is positive the alternative encoding of $-+--$ should be used. So if the input data were 111 111, it would be encoded to $+-++$ $-+--$ or $-+--$ $+-++$, but never $+-++$ $+-++$! If the decoder keeps a record of the running digital sum it will always know what the polarity of the next unbalanced word should be, and any discrepancy will indicate an error. Any single error must change the disparity of a word, so that even if it does not create a forbidden word an error will always be detectable by the discrepancy in the digital sum. The principle can be thought of as the encoder and decoder both keeping a record of the running digital sum between words: in the absence of errors, the record will agree. If an error occurs this will affect the sum at the receiver, but not at the transmitter. The discrepancy between the two will eventually be revealed by a disagreement over the disparity of an unbalanced word. As with AMI, there can be delay between the occurrence of an error and its detection.

Any block code with controlled disparity will allow error detection in this way.

5.3.2   Error detection at a repeater

The methods of error detection discussed above require data to be decoded. This is quite acceptable at a transmission system terminal, where the data is being decoded anyway, but it is often necessary to include error detection at a repeater. To have a decoder at a repeater just for error detection would seem excessive and is in fact unnecessary.

Digital sum variation error detection A common technique used for error detection at a repeater, which avoids the need for decoding, is to use the digital sum variation. This involves an extension of the idea of disparity to a single symbol, rather than a word.

Consider again 3B4B code. The running digital sum is fixed to two levels, say -1 and $+1$, between words. Suppose a count is now kept after the transmission (or receipt at a repeater) of each symbol. As for disparity, the count is $+1$ for a positive pulse, -1 for a negative pulse, but there is

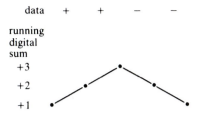

Fig. 5.11. Plotting the running digital sum.

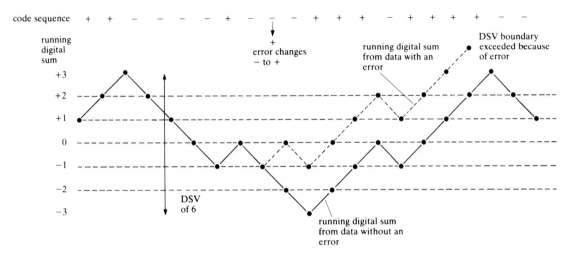

Fig. 5.12. Digital sum variation (DSV) used for error detection.

now a new value of the running digital sum after every symbol. Between words it will always be −1 or +1, but during words it will vary by up to a further +/−2. For example, suppose the running digital sum is +1, and the next code word is the balanced word + + − −. After the first + the running digital sum will be +2, after the second + it will be +3; it then goes back to +2 after the first − and to +1 again after the second −. A useful way of visualising the running digital sum is to plot it as in Fig. 5.11.

Since the running digital sum can be either −1 or +1 between words and can vary by +/−2 during words, the total range it can take is from +3 to −3. The total excursion of the running digital sum is a useful parameter of the code and is called the *digital sum variation* abbreviated to *DSV*. The DSV of this 3B4B code is 3 − (− 3) = 6. A single error will cause the measured running digital sum to exceed the bounds of the code DSV, as illustrated in Fig. 5.12. So error detection is possible by keeping a record of the running digital sum and noting whenever it exceeds the bounds of the code DSV – without needing to know the location of the word boundaries.

EXERCISE 5.5

What is the DSV of
(a) CMI
(b) Manchester code
(c) Manchester bi-phase mark

EXERCISE 5.6

The code sequences below are from a code with a DSV of 3. Which
must contain an error?
(a) $0+--0+--0+--$
(b) $---0+-0+++-+$
(c) $-+++---+--++$

Even mark parity error detection There is also a technique for error
detection without word alignment which can be used for codes, such as
24B1P, which have even parity.

   Fig. 5.13 shows a block diagram of a system using this method. A logic
circuit toggles (changes state) for each received binary 1 (mark); it
remains unchanged for received 0s. The logic levels out from the toggle are
bipolar, and this is low-pass filtered and fed to a comparator. The effect of
the combination of low-pass filtering and applying to a comparator is to
determine the average level of the toggle output: more 1s than 0s and the
comparator output will be a logic 1, more 0s than 1s and the comparator
output will be a logic 0.

   If the received data were completely random, the output from the toggle
would also be random and the mean level would be zero. In practice,
because of its random nature, the toggle output would occasionally
contain long strings of 1s and long strings of 0s. Because of the finite
cut-off frequency of the low-pass filter, these strings of like symbols would
cause the input to the comparator to wander around zero and the
comparator output would alternate in some slow, random, fashion.

   If, however, the data is in blocks which have even parity, the situation is
rather different. Suppose the toggle output starts off high. During the first

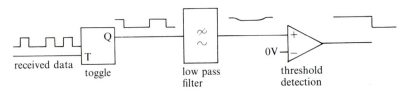

Fig. 5.13. Even mark parity error detection.

block of data it will vary in a random fashion, but, because of the even parity, by the end of the block it will have toggled an even number of times and therefore be back high again. In fact, at the end of each block it will always be high. The fact that it is forced high at the end of each block introduces a slight asymmetry in the toggle output sequence; its mean level will be positive. In the absence of errors, and providing that the low-pass filter is properly designed, the comparator output remains high. If, however, a single error occurs, one block of data will have odd parity. This means that the toggle output will end that block low. If there are no more errors the toggle output will now always be low at the end of data blocks, so the mean level will be low and the comparator output low. In other words, the single error will cause the comparator to change from high to low. If another error occurs, the resultant odd parity block will cause the comparator to go high again.

In summary: the comparator output stays in the same state (be it high or low) in the absence of errors; errors are detected by a change in state of the comparator output.

This technique can be used for a simple, low power, error detector, but it needs careful design and its limitations must be taken into account. For example, the choice of the low-pass cut-off frequency of the filter is critical. If it is too high, the random fluctuations of the toggle output will get through to the comparator and cause spurious 'errors' to be raised. If it is too low, the signal level to the comparator will be very low and therefore susceptible to noise, and the error rate at which saturation occurs will be very low. Furthermore, at very high error rates, rather than the error detector just saturating the apparent error rate can actually fall. An additional 'catastrophic' failure detector may be required.

5.3.3   Decoding data which has been corrupted by errors

If the decoder encounters data which it knows to contain an error, what should it provide as the decoded output? Clearly the aim is to try to ensure that the decoded output contains the minimum number of errors possible. There are a number of criteria which can be used to optimise the output.

Minimum distance criterion   To take an example, consider the decoding of a forbidden word of the 3B4B code: say $- + + +$. Conceptually, the decoder decides which legitimate code word is most likely to have been intended, and provides an output corresponding to the decoding of that legitimate code word. In deciding which code word is most likely to have been intended, the first criterion usually applied is that of *minimum distance*. This criterion says that the best assumption is that which implies the smallest corruption of the data: the intended code word is taken to be the one which has the smallest distance from the received code word. For example, $- + + +$ could have been created by a single error in the second symbol of code word $- - + +$, or it could have been created by two errors, one in the second symbol and one in the fourth, of the code word $- - + -$. The former is taken to be the better assumption. The measure of distance being used here is Hamming distance: just a count of the number of symbols that differ between two code words (a disagreement in disparity would also count as one towards the distance).

The minimum distance criterion indicates that $- - + +$ is a better assumption than $- - + -$ for the word intended when $- + + +$ is received, but $- + + +$ could also be due to $- + - +$ with an error in the third symbol, or $- + + -$ with an error in the fourth symbol. Both these last two have, like $- - + +$, a Hamming distance of 1 from $- + + +$, so the minimum distance criterion cannot help in deciding between the three. There may, nevertheless, be ways of deciding which is the best decoding to use: such as to minimise the so-called error multiplication.

Minimising error multiplication Suppose, continuing the example of the last section, that the decoder has been constructed to assume that $- + - +$ was intended whenever $- + + +$ is received. It therefore gives a decoded output of 010 whenever it receives $- + + +$ (010 is encoded to $- + - +$ in the 3B4B code in Table 5.1). Suppose further that on a particular occasion 011 is transmitted, encoded to $- + + -$, and an error results so that $- + + +$ is received. The net result is that 011 at the transmitter is turned into 010 at the receiver: in other words the single error during transmission has resulted in a single error in the decoded data. On another occasion 001 might be transmitted, encoded to $- - + +$, corrupted by a single error to $- + + +$, and decoded to 010. This time 001 has been converted to 010 by the error in the transmission system. 001 differs from 010 in two places so there are two errors; the single symbol error has created two errors in the decoded data. On yet a further occasion, 010 might be transmitted as $- + - +$, and an error corrupt it to $- + + +$. But this time it will be decoded to 010; the single error in the symbol has not created any errors in the decoded data.

The ratio between the number of errors in the decoded output to the number of errors in the received data is known as the error multiplication or error extension. As illustrated by the above argument about the decoding of $- + + +$ in 3B4B code, the error extension depends upon

   (i) what decoding is chosen for data containing errors; and
   (ii) what the input data is

In analysing a code it is usual to assume that the input data is random. So in the example above, 001, 010 and 011 are equally likely to occur and therefore equally likely to be the source of the forbidden word $-++$. It is therefore useful to take the average of the number of errors in the decoded output and say that when $-++$ is received, there is an average error extension of 1 (0 errors if the input was 010, 1 if the input was 011, 2 if the input was 001; giving an average of 1). This is for the case when the decoder is using 010 for the output from $-++$. The decoding can be analysed in the same way for the case of using either 001 or 011 as the decoding of $-++$. The results for the three possible decodings are summarised in Table 5.8, from which it can be seen that error extension when $-++$ is received is a minimum (of 2/3) if 011 is used as the decoding. The best code performance is therefore achieved by using 011 as the decoding of $-++$: hence the entry for $-++$ in the decoding table (Table 5.3).

With any particular code and decoding table, it is possible to consider the error extension for all possible symbol errors. The average error extension for all possible errors in the received coded data gives the error multiplication factor or error extension of the code. This parameter can be used to predict the error rate in decoded data for a given error rate in received data. For example, the error multiplication factor of the 3B4B code using the decoding table of Table 5.3 is 1.25. This means that if there are R errors per second in the coded data, the error rate in the decoded data will be $1.25 \times R$ errors per second.

Soft decoding There has been an assumption so far that the decoder deals only with purely digital data, which may or may not contain errors. The input to the receiver, however, is an analogue signal from the line which the regenerator has quantised to one of the code signalling states for each received symbol. The decision between different signalling states may have been more marginal in some cases than in others, but this fact is unknown to the decoder. Decoding where the regeneration and decoding are treated as separate functions is known as hard decoding or hard decision decoding. The decision made by the regenerator is 'hard' in the sense that it is one signalling state or the other, there is no 'in-between' (Fig. 5.14).

Some information about how confident the regenerator is in making its decision could be useful to the decoder. Continuing the example from above from 3B4B code, the decoder has to decide whether $-++$ was meant to be $-++-$, $-+-+$ or $--++$. Although the regenerator has found the received word to be $-++$, one of the last three symbols could have been much closer to the threshold for interpretation as a

Table 5.8. *The effect of the choice of decoding for* $- + + +$ *on the error extension*

(a) Decoding $- + + +$ to 011

Input	Transmitted word	Received as	Decoded to	Number of errors
011	$- + + -$	$- + + +$	011	0
010	$- + - +$	$- + + +$	011	1
001	$- - + +$	$- + + +$	011	1
			Average number of errors	2/3

(b) Decoding $- + + +$ to 010

Input	Transmitted word	Received as	Decoded to	Number of errors
011	$- + + -$	$- + + +$	010	1
010	$- + - +$	$- + + +$	010	0
001	$- - + +$	$- + + +$	010	2
			Average number of errors	$3/3 = 1$

(c) Decoding $- + + +$ to 001

Input	Transmitted word	Received as	Decoded to	Number of errors
011	$- + + -$	$- + + +$	001	1
010	$- + - +$	$- + + +$	001	2
001	$- - + +$	$- + + +$	001	0
			Average number of errors	$3/3 = 1$

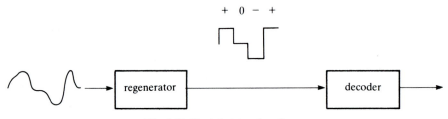

Fig. 5.14. Hard decision decoding.

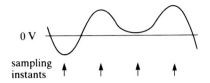

0 V

sampling
instants

Fig. 5.15. Soft decisions.

negative pulse than the other two. Suppose, for example, the input to the regenerator was as in Fig. 5.15. Had the decoder 'known' this, the chance of correct decoding would have been increased by assuming that the code word was $-+-+$.

A process which links decoding and regeneration in this way is known as soft decoding or soft decision decoding.

Practical soft decoding usually takes the form of the regenerator quantising the analogue signal from the line to more levels than just the signalling states. All this information is passed to the decoder, which is clearly much more complex then for hard decision decoding. The measure of distance for the minimum distance criterion takes account of how close each received symbol is to the signalling state thresholds.

Soft decoding has a better chance of correctly decoding a code word perturbed by noise than hard decoding, so for a given decoded error rate a lower line signal to noise ratio can be tolerated than for hard decoding. Typically, for an error rate of 1 in 10^9 it may be possible to work with a signal which is lower by 2 or 3 dB.

Although described here for use with a block code, soft decoding can equally well be used with other coding schemes.

5.4   Timing and synchronisation

5.4.1   Clock recovery

As explained in Section 5.1, one of the purposes of many line codes is to combine the timing information with the data. At a regenerator in a receiver the timing information is then extracted to generate a clock waveform. The clock is used to re-time the data so as to restore, as far as possible, the regular timing intervals of the signal (Fig. 5.16).

Extraction of the timing information (clock extraction) may be possible simply by narrow-band filtering, to give a sinusoidal waveform at the required frequency, followed by a comparator to turn the sinusoidal waveform into a square wave (Fig. 5.17). That, however, would require

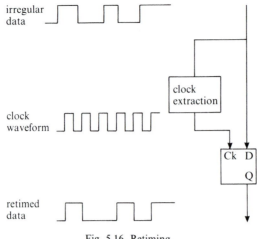

Fig. 5.16. Retiming.

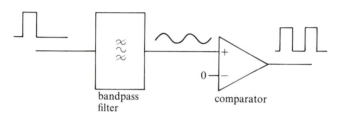

Fig. 5.17. Clock extraction.

the received data to contain a component at the required clock frequency (the data rate), which is rarely the case. More commonly a frequency component at some related frequency is identified and processed in a non-linear way to generate the required clock frequency. In particular, it was emphasised earlier that coded data should have frequent transitions between symbols. A transition between *each* symbol would result in data with a spectrum containing a strong component at half the data rate. Non-linear processing can generate harmonics; including the second harmonic which will be at the required frequency. Less frequent transitions result in components at lower frequencies, but non-linear processing again generates harmonics and the third or higher harmonic will be equal to the required frequency.

Rectification is often the non-linear process used in clock extraction. The complete process is shown in Fig. 5.18: the received data is differentiated to produce a sequence of spikes corresponding to the transitions in the data; full wave rectification makes all the spikes have a positive polarity; a narrow band filter (shown as an *LC* tuned circuit) then

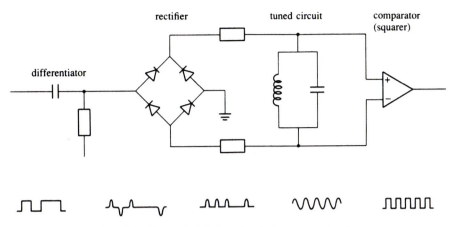

Fig. 5.18. Conceptual design of a clock recovery circuit.

The differentiator is shown in Fig. 5.18 as a *CR* high-pass filter. Provided the cut-off frequency is much higher than the significant frequency components in the signal, a first-order high-pass filter does, in fact, approximate to a differentiator. What is required is that a rectangular wave should be converted into short pulses at the transition instants: a *CR* circuit with sufficiently high cut-off frequency does precisely this, by charging and discharging the capacitor in much less time than the duration of a pulse.

selects the required frequency component; and finally a comparator turns the sine wave into a square wave.

The characteristics of the narrow-band filter are important in the performance of the clock recovery process. Often this filter will be a high-Q (low damping factor) LC tuned circuit as shown in Fig. 5.18, with resonant frequency equal to the required clock frequency. It is useful to consider the action of this filter in the time-domain. Each pulse derived from the transitions in the data can be viewed as an impulse which excites the impulse response of the filter. This impulse response is a decaying sinusoid at the damped natural frequency of the tuned circuit. Superposition can then be used to determine the output of the tuned circuit in response to the sequence of timing pulses. For a filter tuned to exactly the right frequency, and assuming no jitter on the received data, the phases of the decaying sinusoids will be the same for all pulses so that the result of the sequence of pulses is to maintain the oscillation on the output. During gaps in the timing pulses, the oscillation will start to decay, but will be re-enforced on the arrival of the next pulse (Fig. 5.19).

Clock mistuning Although the tuned circuit will have been tuned nominally to the signalling rate, there will in practice be a slight difference (a clock mistuning) resulting from the tolerance allowed on the input signal, errors in originally tuning the clock recovery circuit, and drift (due to aging or temperature changes) in the component values. The natural frequency of the clock recovery circuit will therefore be slightly different from the data signalling rate, and so the phase of oscillations from each impulse will differ slightly. The effect of adding two sinusoids of the same frequency but of different phase is to produce another sinusoid of the same frequency but with some intermediate phase. If, furthermore, one of the

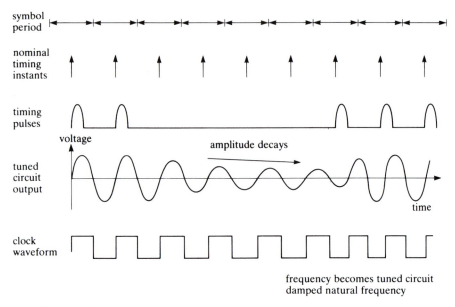

symbol period

nominal timing instants

timing pulses

voltage

amplitude decays

tuned circuit output

time

clock waveform

frequency becomes tuned circuit damped natural frequency

Fig. 5.19. The effect of a gap in the timing information on the clock recovery circuit.

two sinusoids has a larger amplitude than the other the resultant sinusoid will have a phase closer to the one with the larger amplitude. The output from this slightly mistuned clock recovery circuit will therefore be a sinusoid at the tuned circuit natural frequency, with a phase dominated by the impulse response from the latest timing pulse. If the timing pulses are very frequent, the phase will be being constantly updated by the new timing pulses – one way of viewing it is that each timing pulse 'pulls' the oscillations of the tuned circuit to keep them in time with the sequence of timing pulses. This will result in the output being effectively at the rate of the timing pulses – the required clock frequency.

During long gaps in the timing pulses, however, the phase will not be updated so the oscillations will continue at the damped natural frequency. If this frequency is slightly wrong (due to the mistuning), a phase error ϕ will build up on the recovered clock (Fig. 5.20, which greatly exaggerates the rate of decay of the sinusoid). The impulse response of the next timing pulse to arrive will dominate the (now decayed) oscillations, correcting the clock phase by means of a phase step. Before the phase has been corrected, however, problems may have been caused by the phase error. For a small phase error, the retiming of the received data will merely be at a non-ideal instant, resulting in an increased probability of error. The drift followed by the phase step will also add jitter to the retimed waveform. For large drifts, however, the error can become so great that the clock will start to sample in the wrong data period (Fig. 5.21). Furthermore, when a

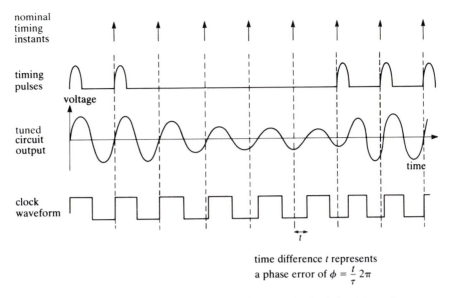

Fig. 5.20. The build-up of phase error during a gap in the timing information.

timing pulse does arrive, although it will correct the phase there will have been one cycle of the clock waveform too few, or one too many, since the previous timing pulse. The overall result is that there has been a slip in the data – a lost symbol or an extra symbol gained. Such a slip (a 'bit slip') is very serious because it means that any subsequent decoders will need to realign and until this has happened the output data will be completely useless.

Jitter reduction The retiming process can add jitter to a waveform due to clock mistuning, but the clock recovery circuit will also, to a greater or lesser extent, reduce jitter. Qualitatively, this is because the clock recovery circuit will be unable to follow very rapid timing irregularities. The effect is for the clock recovery circuit to act as a low-pass filter to the jitter waveform. Quantitatively, the jitter filtering effect of a tuned-circuit clock recovery has a first-order characteristic with low-pass cut-off frequency of $f_0/2Q$, where f_0 is the signalling rate of the recovered clock, and Q is the Q-factor of the clock recovery circuit. In other words, a clock recovery circuit with a high Q-factor will remove lower frequency jitter than one with a lower Q-factor.

In general, therefore, it is desirable to aim for the highest practicable Q factor in a clock recovery circuit, in order to remove as much jitter as possible. Very high Q-factors can be achieved by the use of phase locked loop clock recovery circuits (see, for example The Open University, 1990).

Virtually any processing of the signal will tend to add jitter. For example, imperfect threshold detection might result in the data transitions occurring at the wrong time, adding jitter.

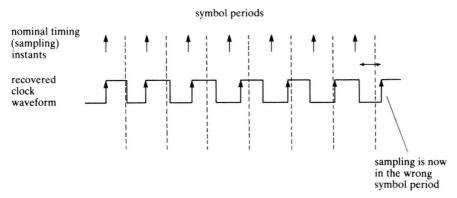

Fig. 5.21. Bit slips resulting from clock mistuning.

Since a clock recovery circuit must follow the signalling rate of the incoming data, however, it must always allow very low frequency jitter (wander) through unattenuated.

5.4.2   Block alignment

The timing and clock recovery discussed above are concerned with symbol synchronisation; that is, ensuring that the signalling rate clock at the receiver is synchronised to the received signalling rate. For block codes an additional synchronisation is necessary: the receiver must know where to break the incoming data into blocks. This is word or block synchronisation, otherwise known as block alignment.

Consider the following CMI encoded data:

$$\ldots + - + + + - - - + \ldots$$

Trying the two possible ways of breaking the data into pairs of symbols (the two possible 'phases') gives $\ldots + - + + + - - - + \ldots$ or $\ldots + - +$ $+ + - - - + \ldots$. From a knowledge of CMI it is clear the second of these must be the correct way to break the data, since the first generates pair of the form $+ -$ which are not valid code words.

On the other hand, consider the following Manchester coded data:

$$\ldots + - + - + - + - + \ldots$$

Trying the two possible phases in this case gives $\ldots + - + - + - + -$ $+ \ldots$ or $\ldots + - + - + - + - + \ldots$. Either of these *could* be correct, and since they would be decoded to different data it is important to know which is the one that should be used. Unfortunately, with this sequence in isolation, it is impossible for the receiver to decide which is the correct phase. However, once the data starts to vary, one of the two phases would produce words that are not part of the coding table, allowing the rejection of that phase. Suppose, for example, it continues thus:

$$\ldots + - + - + - + - + + - - + \ldots$$

The two phases would be either

$$\ldots + \ \ - + \ \ - + \ \ - + \ \ - + \ \ + - \ \ - + \ldots$$

or

$$\ldots + - \ \ + - \ \ + - \ \ + - \ \ + + \ \ - - \ \ + \ldots$$

Since the latter results in words $+ +$ and $- -$ appearing, which are not legitimate code words, the receiver can reject that option.

The Manchester code example above shows that there are data patterns to which the receiver cannot align (once the receiver is aligned, of course, in the absence of any degradation it will remain aligned). This is a common characteristic of block codes (CMI being one exception – which is one of its advantages) and is overcome by ensuring that the data does not continue indefinitely in one pattern. The problem can be ignored if it can be assumed that the input data will contain sufficient variations; otherwise some additional coding may be required. Scrambling (see Chapter 6) is sometimes used for this purpose.

The argument above assumes that the data is error-free. The situation is slightly more complicated in a practical system where no such assumption is possible, since errors themselves can create 'forbidden words', making the receiver think it is in the wrong phase when it is in fact correct. To reduce the probability of the receiver spuriously changing alignment due to errors, the block alignment control must employ some strategy to distinguish between the two sources of forbidden words. One possibility is to change to another phase only if substantial improvement results. A simpler method is merely to require a relatively large proportion of forbidden words before changing phase. Spurious phase changes would then occur only in the event of a particularly high error rate.

5.5 Spectral considerations

The spectra of individual pulses were considered in Chapter 4. We now want to look at the influence of line codes upon the spectrum of transmitted data. A convenient way to do this, using our layered model, is based on the spectrum of the sequence of impulses passed from the line code layer to the pulse layer. The advantage of using the layered model in this way is that it separates the effects of the code from the effects of the pulse shape. Because pulse generation is a linear process, the spectrum of the transmitted signal is obtained by multiplying the spectrum of the impulse train (determined by the line code) by the frequency response of the pulse generator (which is equivalent to the pulse spectrum).

When comparing various codes and pulses, it is usual to assume that the data input is random, in which case (as discussed in Chapter 3) it is appropriate to use power spectra and consequently the relevant input–output relationship is

$$S_o(f) = S_{lc}(f) \times |G(f)|^2$$

where $S_o(f)$ is the power spectrum of the output pulse train, $S_{lc}(f)$ is the power spectrum of the impulses from the code layer (the *code spectrum*) and $G(f)$ is the frequency response of the pulse generator (the pulse spectrum).

Notice that uncoded bipolar data will have a flat 'code spectrum'. This is because the output from the line coding layer will be a random sequence of bipolar impulses, which, as noted in Section 3.4.3, has a constant, frequency independent, power spectrum. In other words, since the impulse train corresponding to bipolar random data is flat, the power spectrum of *uncoded bipolar* data is determined solely by the spectrum of the pulses used. When line coding is employed, however, the impulse train will have a spectrum which is no longer flat, and the output of the pulse generator will be affected by both the pulse spectrum and the spectral characteristics of the line code $S_{lc}(f)$. We shall now look at spectral features of $S_{lc}(f)$ for some of the codes discussed earlier.

5.5.1   Code spectra

Fig. 5.22 shows the spectra (power spectral densities) of a number of the codes which were discussed earlier in this chapter. There are a number of important points to note. Firstly, the spectra have been drawn normalised to the signalling rate. For some comparisons it might be more useful to draw them normalised to the data rate; this would alter the relative horizontal scaling considerably. For example, for a given data rate, CMI (which has been analysed as a 1B2B code) would have twice the signalling rate of AMI, so the AMI spectrum should be compressed to half the width of the CMI spectrum. Another important feature is that the spectra repeat at frequency intervals equal to the signalling rate (only the beginning of the first repetition is shown in the figure). This is rather like the spectrum

Assuming a random data input is convenient for general comparisons – it enables a single code spectrum to be presented for each code – but the code spectrum so derived needs to be used with care. The input data may sometimes have a significant degree of structure; it could, for example, run for long periods with all data 0s. Under those circumstances the spectrum of the coded data may bear little resemblance to the random data code spectrum.

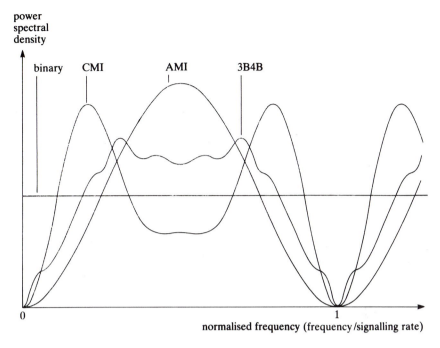

power
spectral
density

binary CMI AMI 3B4B

0 1

normalised frequency (frequency /signalling rate)

Fig. 5.22. Power spectral densities of various line codes.

of a sampled signal as discussed in Chapter 2 – which is not surprising since the sequence of impulses from the line code layer is very similar to our model of a sequence of samples, with the 'sampling' frequency equal to the signalling rate. Remember, however, that the spectrum of an individual pulse is not repetitive; it rolls-off with frequency at a rate which depends upon the pulse shape. So although the code spectrum is repetitive, the spectrum of the coded pulses is not.

5.5.2 Baseline wander

Looking at code spectra can be useful for making a qualitative comparison between line codes, but in this section we look at a method of quantitatively assessing the impact of a code's low frequency characteristics.

It has been mentioned several times that one of the reasons for using a line code is to allow data to be conveyed over a channel with zero d.c. response (d.c. blocking). Typically, such a channel would be one where a d.c. blocking capacitor has been put in series with the line to prevent problems with different earth voltages at different locations. The input circuit at the receiver of such a line would then look like Fig. 5.23. The

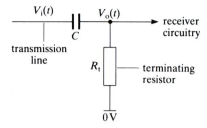

Fig. 5.23. The input circuit of a receiver with d.c. isolation.

resistor is required to terminate the transmission line correctly, and the resistance value will therefore be determined by the nature of the transmission line.

At first sight it might seem that, providing the low frequency cut-off of the high-pass filter formed by the combination of R_t and C is very much lower than the signalling rate, the capacitor will be effectively transparent to any data. This is not necessarily so, however, because even data with a very high signalling rate can have low frequency components for some data patterns. The result is baseline wander when the capacitor starts to build up charge due to short-term imbalances in the number of positive and negative pulses.

To quantify the effect, suppose that the signal uses pulses which have amplitude $\pm A$ and period τ. $V_i(t)$ is used to describe the waveform before the capacitor and $V_o(t)$ to describe the waveform after the capacitor. Whenever there is a positive pulse, the capacitor will be charging up 'positively' – i.e. so as to reduce $V_o(t)$ compared to $V_i(t)$. Whenever there is a negative pulse, the capacitor will be discharging (or charging 'negatively'). Assuming that the voltage offsets introduced by this effect are small, the amount of charge added to the capacitor will be the same for each positive pulse (say q), and will be equal and opposite for each negative pulse ($-q$). After equal numbers of positive and negative pulses therefore there will be no charge on the capacitor, and $V_o(t)$ will be equal to $V_i(t)$. If at some time there have been n more positive pulses than negative, then the capacitor will contain a charge of nq.

Recall that the count of the difference between the number of positive and negative pulses is called the running digital sum. So the charge on the capacitor at the end of each pulse varies with the running digital sum, scaled by a factor q.

An estimate of the value of q may be determined quite simply. During the pulse of voltage A a current of approximately A/R amps will flow (Fig. 5.24). (Any voltage offset across the capacitor is small – specifically it is very much less than A – so the voltage across the resistor is approximately

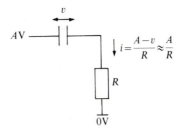

Fig. 5.24. Current during a pulse of voltage A volts.

A). So during the pulse of period τ a total charge of $q = A\tau/R$ will flow into the capacitor.

The voltage offset caused by a charge q on the capacitor may also be easily calculated: it is simply $\delta v = q/C$.

Combining these results, the voltage offset across the capacitor at the end of each pulse is given by

$$v = n\delta v = nq/C = nA\tau/C$$

Where n is the current value of the running digital sum. This effect is shown in Figs. 5.25 and 5.26 (for a binary and ternary code respectively). The baseline wander can be clearly seen to be plotting out the (negative) running digital sum.

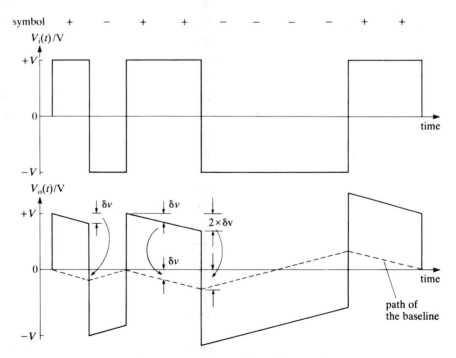

Fig. 5.25. Baseline wander in a binary code.

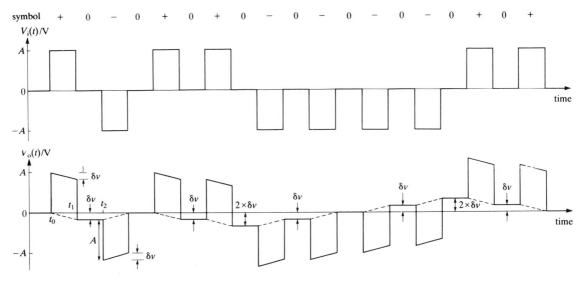

Fig. 5.26. Baseline wander in a ternary code.

Of particular interest is the digital sum variation (DSV) of a code. This gives the maximum excursion of the running digital sum. The maximum peak-to-peak excursion of the baseline wander is therefore given by

$$V = \text{DSV} \times A\tau/C$$

This gives a quantifiable measure of the effect of the capacitor on the data waveform. It can be seen, for example, that increasing C decreases the baseline wander.

The effect on the receiver's performance can best be seen with the aid of an eye diagram. Eye diagrams for the data of Figs. 5.25 and 5.26 are shown in Figs. 5.27 and 5.28. In a real system, of course, you would never see such sharply defined eyes. There would always be noise 'blurring the edges' and low-pass filtering would have led to the corners being rounded. However, Figs. 5.27 and 5.28 can be used to estimate the degradation caused by baseline wander in isolation. Baseline wander appears on the eye diagram as more than one level for each state. This multiplication of levels reduces the gap that separates the levels representing different states. A reduction in this gap is a reduction in the 'safety margin' for distinguishing states: it increases the probability of errors being caused by noise. The gap is known as the eye opening and a reduction in the gap is referred to as eye closure.

In the case of the ternary code (Fig. 5.28), the eye opening in the absence of baseline wander would be AV for each pulse polarity. Baseline wander has reduced this on the right hand side of Fig. 5.28 to $(A - 4 \times \delta v)V$ (the

Eye diagrams were discussed in Chapter 4.

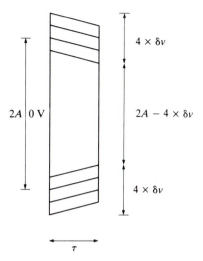

Fig. 5.27. Eye diagram for the data in Fig. 5.25.

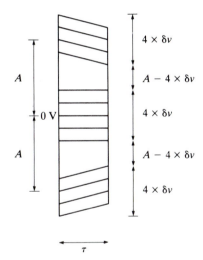

Fig. 5.28. Eye diagram for the data in Fig. 5.26.

right hand side is the worst case position). The fractional eye closure is therefore $4 \times \delta v/A$. In general, for a ternary code with DSV equal to D, the fractional eye closure is:

$$D \times \delta v/A$$

or, substituting for δv

$$\text{fractional eye closure (ternary code)} = \frac{D\tau}{R_t C}$$

The situation is slightly different for a binary code. The eye opening

Table 5.9 *2B3B code*

Input	Code words		
	A	B	C
00		$--+$	
01		$-+-$	
10		$+--$	
11	$+-+$		$---$

Word A has disparity $+1$, words B have 0 disparity and word C has -1 disparity (see text).

without baseline wander in this case would be $2A$, so the fractional eye closure (for Fig. 5.27) is $4 \times \delta v/2A$. In general

$$\text{fractional eye closure (binary code)} = \frac{D\tau}{2R_tC}$$

An alternative way of writing these expressions is to substitute R_tC by $1/2\pi f_c$ where f_c is the low frequency cut-off of the RC combination viewed as a high pass filter. This gives

$$\text{fractional eye closure (ternary code)} = 2\pi f_c D\tau$$

and

$$\text{fractional eye closure (binary code)} = \pi f_c D\tau$$

It has been implicitly assumed in this argument that full-width pulses are being used, with τ equal to 1/signalling rate. The argument does not, however, require that assumption. The formulae are valid for any pulse width, provided τ represents the actual pulse width (rather than the signalling interval).

Notice also that the eye closure is caused by the *variation* in the voltage offset across the capacitor. A *fixed* offset would come about if the coded data (as transmitted) contained a constant d.c. component. Such an offset would not cause eye closure – but it is usually still undesirable because it could cause difficulties with the setting of threshold detection in a regenerator. An example of a code with a fixed offset is the 2B3B code with the coding table of Table 5.9. After passing through a channel with d.c. blocking, this code would have a fixed offset equal to 1/6 of the peak-to-peak amplitude (so positive pulses would have amplitude of $^2/_3A$ and negative pulses would have amplitude $^1/_3A$, where A is the peak-to-peak amplitude, Fig. 5.29). The analysis of the code is slightly

Consistency can be achieved by counting 1/2, rather than 1, for each pulse in a binary code – then the same formula for fractional eye closure may be used for all cases (i.e. the formula for a ternary code).

Table 5.10. *An alternative 2B3B code*

Input	Code word
00	$- - +$
01	$- + -$
10	$+ + -$
11	$+ - +$

more complex in this case, but one approach is to count each negative pulse as 1/3 and each positive as 2/3. The code DSV is then $2^1/_3$, and the baseline wander is given by the formula for eye closure of a ternary code.

Even a code in which the 'd.c.' component is variable may sometimes be usable on a channel with d.c. blocking. Table 5.10 is another 2B3B code; one for which the d.c. offset depends upon the input data. If the input only contains the words 00 and 01, then the output will have a positive offset of 1/6 of the peak-to-peak amplitude (as with the previous 2B3B code). If the input data only contains the words 10 and 11, then the output will have the same offset, but in the opposite direction. Any combination of the two sets of words will lead to an offset between the two extremes. In the presence of random input data, the output will wander between the two extremes and close the data eye. The extent of the eye closure can be seen from Fig. 5.30 to be 33%. Notice that for this code increasing the capacitor value (in a capacitively coupled channel) will not, in general, decrease the amplitude of the baseline wander; merely reduce the wandering rate. The Fibre Distributed Data Interface (FDDI), a high speed optical fibre local area network, uses a code of this type constructed by 4B5B coding followed by differential coding. The choice of code alphabet results in a baseline wander of 10% (Hamstra & Moulton); this is tolerated in FDDI because of other benefits of the code. For example,

33% is the minimum eye closure that can be achieved in the limit of high capacitance. If the capacitor is too small, an additional baseline wander will be experienced above and below the two fixed offsets.

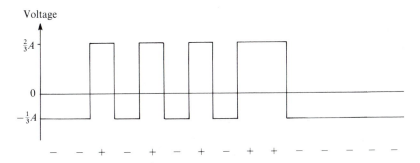

Fig. 5.29. 2B3B code waveform.

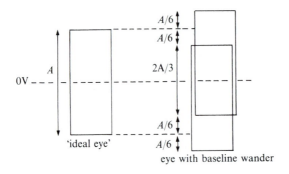

Fig. 5.30. Baseline wander for the 2B3B code of Table 5.10.

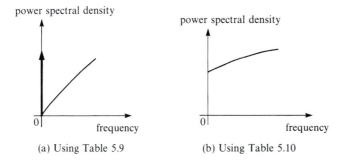

(a) Using Table 5.9

(b) Using Table 5.10

Fig. 5.31. Code spectra for two different 2B3B codes.

coding of input bits in blocks of 4 allows the line coding to be synchronised to the byte (8-bit) structure of the carried data.

Baseline wander and code DSV can be related, qualitatively, to the appearance of code spectra at the low frequency end. Looking back at Fig. 5.22, you can see that AMI, which has the lowest DSV (1), has the smallest low-frequency component (its spectrum falls off most rapidly as it approaches d.c.). At the other extreme, binary bipolar NRZ has unlimited DSV and its spectrum does not fall off at low frequency at all. In between these two extremes are CMI and 3B4B with DSVs of 3 and 6 respectively.

The spectra of the two 2B3B codes near d.c. are shown in Fig. 5.31. The code which uses Table 5.9 has a spectrum which falls off to zero near d.c. (because the offset is fixed), but has a spectral line (mathematically a frequency domain impulse) at d.c., corresponding to the finite d.c. power. The code which uses Table 5.10, in contrast, does not fall off to zero towards d.c. (because its offset is not fixed), but it does not have a spike at d.c. because its long-term mean voltage is zero.

EXERCISE 5.7

What do the DSVs of CMI and Manchester code indicate about their tendency to baseline wander? If a 5% eye closure due to baseline wander is acceptable, what high-pass cut-off frequency is acceptable in each case for a line system conveying data at a rate of 140 Mbits s^{-1}?

5.6 Summary

The main functions of most line codes are:

1. To allow a baseband signal to be conveyed over a channel with d.c. blocking.
2. To ensure that the coded data contains enough timing information for a clock waveform to be extracted at the receiver.
3. To allow error detection to be performed.

Line coding is usually binary or ternary, and timing information is combined with the data by ensuring that the code sequence contains sufficiently frequent transitions. The amount of timing information contained in the code sequence – measured by the length of gaps with no transitions – can be balanced against the sophistication of the clock recovery circuits in the regenerators used by the transmission system. Very low timing content (such may be found when using 24B1P or 2B1Q codes) can be tolerated by using clock recovery circuits with very high Q-factors. Insufficient timing content in the code will cause jitter on the data, increased susceptibility to errors and, in the extreme, bit slips.

Low frequency components in the code spectrum are kept down by ensuring that positive pulses are balanced by negative pulses over a reasonably short period. Excessive low frequency in a code causes baseline wander if the data is transmitted over a channel with zero d.c. response. A measure of the code's tendency to cause baseline wander is given by examining the code spectrum or by the code's digital sum variation (DSV).

For balanced codes, the DSV can also be used for error monitoring purposes by checking if the measured running digital sum exceeds the DSV: this happens only in the event of an error. Alternative error monitoring methods are possible at a decoder by checking if the received data violates the coding rules.

Decoding aims to minimise the number of errors in the output data even in the presence of errors in the received data. The ratio of the number

Table 5.11

| Code parameters | | | | | Bandwidth efficiency (bits Hz^{-1}) | | |
Code	Radix	Efficiency (%)	Timing	DSV	Nyquist filtering	Raised cosine 50%	100%
Uncoded binary	2	100	UB	UB	2	1.33	1
24B1P	2	96	UB	UB	1.92	1.28	0.96
3B4B	2	75	4	6	1.5	1	0.75
CMI	2	50	3	3	1	0.67	0.5
Manchester	2	50	2	2	1	0.67	0.5
AMI	3	63	UB	1	2	1.33	1
HDB3	3	63	3	2	2	1.33	1
4B3T	3	86	4	7	2.67	1.78	1.33
2B1Q	4	100	UB	UB	4	2.67	2.00

Partial response signalling

Code parameters (see note 1)

Scheme	Radix	Efficiency (%)	Timing	DSV	Bandwidth efficiency
Duobinary	3	63	UB	UB	2
Lender's modified duobinary	3	63	UB	2	2

Notes:
1. The code parameters for the partial response schemes are derived from consideration of the symbol sequence detected at the receiver. They can be used for comparisons with other codes, but should be treated with caution.
2. The figures for timing content refer to the maximum possible run of symbol intervals which can occur without a voltage transition. 'UB' (unbounded) indicates that indefinite runs without transitions are possible.
3. UB (unbounded) for the code radix implies that the running digital sum can increase or decrease without limit.

of errors in the decoded data to the number of errors in the received coded data is known as the error extension. Error extension is minimised by assuming minimum distance between the intended and received waveform. Soft decoding can give a lower error ratio than hard decoding but is more complex to implement.

The choice of a code for a particular application will involve a trade-off between the need to keep the coding circuits as simple as possible and the need to achieve the desired characteristics in the code sequence. Shorter links call for simpler coding circuits because the higher cost of more complex circuits is less worthwhile. This often leads to the use of AMI or related codes such as HDB3. Codes for long transmission systems are often more complex – and more efficient – block codes. The use of a more efficient code results in a lower signalling rate for a given data rate, and consequently allows longer distances between repeaters.

The parameters of some of the codes which have been described in this chapter are compared in Table 5.11, together with a few new ones. The parameter labelled 'bandwidth efficiency' is a measure of the combination of line coding and pulse shaping, and is calculated from the ratio of the conveyed data rate (in bits per second) to the required channel bandwidth (in Hz). For example, the theoretical combination of an uncoded bipolar signal using a Nyquist channel would require a channel bandwidth of $B/2$ Hz to convey a data rate of $1/B$ bits per second, giving a bandwidth efficiency of 2 bits Hz^{-1}.

Duobinary and Lender's modified duobinary can be considered as combined coding and pulse generation schemes, so have been included in the Table. The code parameters given for these are for the pulse sequence as interpreted by the receiver, following filtering by the channel.

6

Channel codes

6.1 Introduction

In the Introduction to Part 2, the channel coding layer was identified as being concerned with maintaining the integrity of the conveyed data sequence. In one sense, of course, both line coding and pulse shaping are also chosen with this ultimate end in mind. The precise form of channel coding employed in a particular application depends therefore on the characteristics of the data, the nature of the channel, and the choice of line code and/or pulse shape.

The term channel coding is often used as a near-synonym for error detection/correction coding. Examples discussed in this chapter are Hamming codes, cyclic redundancy check (CRC), and convolutional coding. We have chosen a rather wider interpretation of the term channel coding, however, so as to include techniques sometimes used to make up for limitations inherent in a particular line code or signalling scheme. For example, the line codes HDB3 and CMI guarantee a good timing content, whereas AMI does not. Systems using AMI therefore sometimes incorporate scrambling before line coding to reduce the probability of long strings of data without transitions. Similarly, differential coding can be used to overcome the problems of error propagation which can be encountered with partial response signalling.

There is also another reason for treating together such topics as differential coding, scrambling, CRC and convolutional coding. As will be seen in what follows, the implementation of all these processes has much in common, particularly in the use of delay elements, feedback and shift registers. Differential coding is a particularly simple example and hence a suitable starting point.

6.2 Differential coding

Differential coding, as its name indicates, produces an output in which the information is contained in differences between successive symbols.

The coding rule is generally as follows:

Table 6.1. *The truth table of an exclusive-OR gate*

A	B	$C = A$ ex-OR B
0	0	0
0	1	1
1	0	1
1	1	0

if the input bit is a 1, the output changes state

if the input bit as a 0, the output remains the same

Consider encoding the input sequence 110010. There are two possibilities, depending upon whether the previous output was a 1 or a 0.

Input:	1 1 0 0 1 0
Output (previous output was a 1):	(1) 0 1 1 1 0 0
Output (previous output was a 0):	(0) 1 0 0 0 1 1

Note that the two possibilities are complements of each other.

The block diagram of a circuit to perform differential coding is shown in Fig. 6.1. The storage element would be a D type flip-flop, clocked by the clock waveform associated with the data. The clock has not been shown explicitly: the important feature of the storage element is just that it holds one bit or symbol for the duration of one bit/symbol interval.

The previous output symbol is held in the storage element, and this is compared with the new input symbol in an exclusive-OR logic gate. The truth table of an exclusive-OR gate is shown in Table 6.1, and a few minutes consideration of this table and Fig. 6.1 should reveal how the exclusive-OR gate implements differential coding as described above.

Notice that at the start of the coding of the sequence the previous output symbol is held in the storage element. So whether the coded output

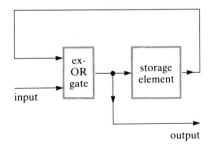

Fig. 6.1. Differential encoder.

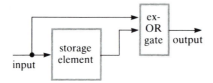

Fig. 6.2. Differential decoder.

is one sequence or its complement depends upon the initial state of this storage element.

Decoding is performed by providing an output 1 if two successive symbols differ, and a 0 if two successive symbols are the same. A block diagram for a decoder is shown in Fig. 6.2.

The decoder does not need to 'know' which of the two alternative possible sequences it is decoding. Since it just looks for differences between successive symbols, the question as to whether the data is complemented or not is irrelevant. This also means that it does not matter if the data is inverted in the channel between encoder and decoder.

Fig. 6.3 shows an encoder and decoder connected together. Imagine the system being switched on and then data being passed through this combination bit by bit. Assume that when the system is switched on, the value taken by the storage elements is 0 or 1 at random. The encoding of the first input bit depends upon the value of the storage element at the encoder. When this first code symbol reaches the decoder, the value to which it is decoded depends upon the value of the storage element at the decoder. If the values of storage elements at the encoder and decoder are

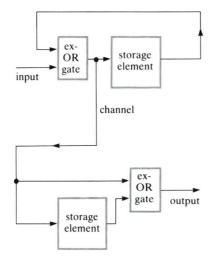

Fig. 6.3. Differential encoder and decoder linked.

the same, then the decoding of this first symbol will be correct. If, on the other hand, the start-up values stored at the encoder and decoder differ, then the first conveyed symbol will be decoded incorrectly.

This first code symbol, however, will be stored in the storage elements at both the encoder and the decoder. So when the next bit is encoded, the storage elements at the encoder and decoder will contain the same value, and the decoding will be correct. From now on in fact (excepting errors occurring in the system) the storage elements at encoder and decoder will always agree and transmission will be correct.

The way in which the system synchronises without any need for external influence, even though an error may occur on start-up, leads to differential coding being described as *self-synchronising*.

If a single code symbol is inverted due, for example, to noise, it will cause two errors in the decoded output at the receiver because it is involved in the decoding of two bits. It causes one error when it arrives at the decoder, then it is stored in the storage element at the decoder and causes an error in the next decoded bit. After that it has no effect on the data.

Differential coding is not generally used to eliminate undesirable data patterns, but other non-redundant coding is. Scrambling, for example, (discussed below) can be used to eliminate long strings of 0s or 1s when there are certain restrictions on the input sequences.

In differential encoding, for every input bit there is a single binary output symbol. Consequently, differential coding is non-redundant (redundancy $= 0$, efficiency $= 100\%$). Being non-redundant has a number of consequences. Firstly, error detection is not possible. For, although a coding rule is followed, all possible binary sequences are legitimate code sequences; it is never possible to say that a given code sequence violates the coding rule.

Secondly, because all possible binary sequences are possible code sequences, non-redundant coding cannot guarantee the elimination of undesirable data patterns – unless there is a restriction on the input data.

An advantage of a non-redundant coding, of course, is that it *is* non-redundant: it introduces no overheads.

6.2.1 Applications

Differential phase shift keying One important application of differential coding is with phase shift keying (PSK). It was explained in Chapter 4 that coherent detection relies on being able to extract a carrier, in the correct phase, at the receiver. Although a carrier *can* be extracted, using non-linear processing, for PSK, uncertainty remains as to whether it is in-phase or anti-phase. If the receiver gets the reference phase wrong, the decoded data will be inverted. This ceases to be a problem if the data has been differentially encoded by the associated channel coding, because, as

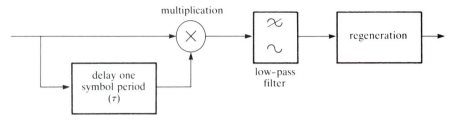

Fig. 6.4. Differentially synchronous detection of DPSK.

explained above, there is no need to know the absolute polarity of differentially encoded data.

Often the differential coding and PSK are combined and treated as a single form of keying, referred to as differential phase shift keying (DPSK). The combined scheme could then be described by, for example:

Input data	Coding
1	phase change of 180°
0	no phase change

EXERCISE 6.1

Show that the encoding of 1100101011 by DPSK gives the same output as differential coding followed by PSK.

The demodulation and decoding can also be combined at the receiver by comparing the phase of successive symbols. A particularly simple receiver (Fig. 6.4) is then possible because, by multiplying successive symbols, the need for carrier extraction is eliminated. As shown in Fig. 6.5, if there is a change of phase between successive symbols the result of the multiplication is a waveform with a positive mean voltage; if there is a phase change of 180° the result of multiplication is a waveform with a negative mean voltage. This type of detection is sometimes described as *differentially synchronous detection*.

There is a slight degradation in performance for DPSK compared to PSK because of the error multiplication of differential coding. With differentially synchronous detection, the differential code error extension does not apply directly, because the decoding is not independent of the detection. There is an equivalent degradation, however, which arises from the fact that both the symbols multiplied together will be contaminated by received noise. In ideal correlation detection the locally generated waveform would be noise free; only the received symbol carries the noise.

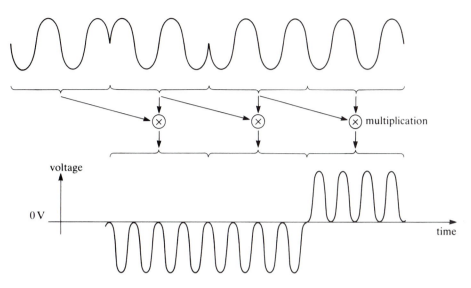

Fig. 6.5. The waveforms of differentially synchronous detection of DPSK.

Despite this degradation, DPSK with differentially synchronous detection is a very useful signalling system because it is relatively easy to implement and offers performance which compares favourably with other, more complex, schemes.

Partial response signalling Another use of differential coding is for the prevention of error propagation in partial response signalling.

Error propagation can occur in partial response signalling because the decoding of some received symbols depends upon the result of previous decoding. Thus, if one error occurs, it can affect the decoding of subsequent symbols.

In the case of duobinary, on receipt of a voltage near zero the receiver output is 'the complement of the previous output'. Thus an error near the beginning of a string of '0V' symbols leads to the decoding of the whole string incorrectly. An example of this is shown in Table 6.2, where the second symbol should have been − V but was incorrectly detected as 0. In this example comparing the input with the incorrect output shows that the single decoding error has led to 10 errors in the decoded output.

Differential channel coding again overcomes this problem. Compare the input data with the decoded data in Table 6.2. The error propagation has inverted the data for the duration of the run of 0Vs. Since being inverted is not a problem with differentially coded data, had this been differentially coded data errors would only have occurred at the start and finish of the run. Table 6.3 shows the case for differentially coded data.

Table 6.2

input data	0	0	0	1	0	1	0	1	0	1	0	1	1	1
Duobinary sequence		−V	−V	0	0	0	0	0	0	0	0	0	+V	+V
			:											
Received duobinary		−V	0	0	0	0	0	0	0	0	0	0	+V	+V
Decoded data	0	0	1	0	1	0	1	0	1	0	1	0	1	1

Table 6.3

Input data		0	0	1	1	1	1	1	1	1	1	1	0	0
Differentially coded	0	0	0	1	0	1	0	1	0	1	0	1	1	1
Duobinary sequence		−V	−V	0	0	0	0	0	0	0	0	0	+V	+V
			:											
Received duobinary		−V	0	0	0	0	0	0	0	0	0	0	+V	+V
Duobinary decoding	0	0	1	0	1	0	1	0	1	0	1	0	1	1
Differentially decoded data		0	1	1	1	1	1	1	1	1	1	1	1	0

Comparing the input data with the differentially decoded data in Table 6.3 reveals only 2 errors. Furthermore, a significant simplification is possible by combining the duobinary decoding with the differential decoding. If, in Table 6.3, you compare the input data with the duobinary sequence, you will notice that input 1s correspond to 0V, and input 0s correspond to either $+V$ or $-V$. This must always be the case, as can be seen from the following:

Input		1	
differential coding	$0 \rightarrow 1$	or	$1 \rightarrow 0$
received under duobinary scheme	0V		0V
Input		**0**	
differential coding	$1 \rightarrow 1$	or	$0 \rightarrow 0$
received under duobinary scheme	$+V$		$-V$

Thus the decoding of the differential duobinary scheme merely requires $+V$ and $-V$ to be decoded to 0, and 0V to be decoded to 1. Using this strategy in Table 6.3, you will find that it not only works, but removes one of the errors.

Other partial response schemes require different coding to prevent error propagation. For example Lender's modified duobinary, referred to in Chapter 4, requires a variation of differential coding in which the input bit controls whether the output symbol is the same as, or the complement of, the last but one output symbol. Thus the encoder required would be similar to Fig. 6.1, but there would be two stages of storage elements to delay the output symbol by two periods before exclusive-ORing it with the input.

6.3   Scrambling

The description 'scrambler' is given to many coding processes which in some way randomise data. This section contains descriptions of two common types of scramblers: 'self-synchronising scramblers' and 'set–reset scramblers'.

6.3.1   Self-synchronising scramblers

This type of scrambling is an extension of the idea of differential coding. Instead of just a single storage element, scrambling makes use of several storage elements formed into a shift register. This extra complexity means that a greater degree of randomness can be created in the data. Scrambling can randomise not only strings of all 0s and all 1s, but also short repetitive sequences. It is conveniently described by looking at the construction of the encoding and the decoding circuits (the scrambler and the descrambler).

Fig. 6.6 is a block diagram of a typical scrambler. (The storage elements are labelled stage 1 to stage n, and again they would be D type flip-flops clocked by the clock waveform, which is not shown). The output from the

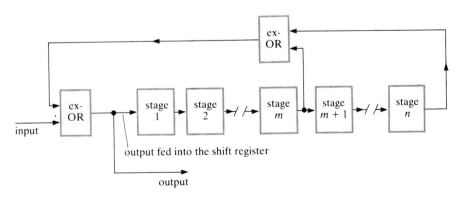

Fig. 6.6. An n-stage scrambler.

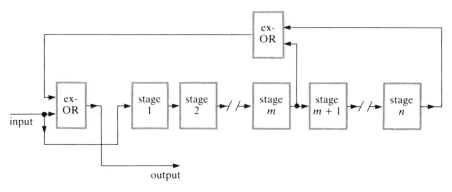

Fig. 6.7. An *n*-stage descrambler.

scrambler is fed into a shift register with *n* stages, so the scrambler contains a copy of the previous *n* coded symbols. From some point along the shift register (from stage *m*) and from the end, feedback taps are taken and exclusive-ORed together. The result of this exclusive-OR is exclusive-ORed with the input to provide the scrambler output.

Fig. 6.7 shows the corresponding descrambler. This is very similar to the scrambler, but notice that the shift register is here fed directly from the input rather than the output as for a scrambler.

Now consider the nature of the encoding performed by the scrambler. The encoding of each input bit (to a 1 or a 0) depends upon the value of the input bit and the feedback taps. Since the feedback is taken from the shift register containing previous code symbols, the encoding of each input bit depends upon the encoding of previous bits. The encoding of any input sequence cannot therefore be taken in isolation: rather the encoding of a

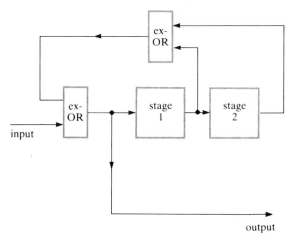

Fig. 6.8. A 2-stage scrambler.

Table 6.4. *Encoding of a data sequence by a 2-stage scrambler*

Input bit	Stage 1	Stage 2	Stage 1 ex-OR Stage 2	Output
1	0	0	0	1
1	1	0	1	0
0	0	1	1	1
0	1	0	1	1
1	1	1	0	1
1	1	1	0	1
0	1	1	0	0
0	0	1	1	1
1	1	0	1	0
1	0	1	1	0
0	0	0	0	0
0	0	0	0	0
1	0	0	0	1

The last line is the same as the first, so the output will repeat the first 12 lines giving 101111010000101111010000....

scrambler is a continuous process beginning when the device is switched on. Furthermore, the code sequence will depend upon the starting state of the scrambler: i.e. the pattern of symbols in the shift register when it is switched on. For an *n*-stage scrambler, there are 2^n possible starting states.

Fig. 6.8 is a block diagram of the simplest example of a scrambler – one with just two stages. The output for any given input can be determined by means of a table. Table 6.4 works through the scrambling of a repeating four bit input pattern 11001100..., for the case of the scrambler starting off with 0 in both of the stages. The last line in this table is the same as the first, and occurs at the same point in the input pattern (the first 1 of the two 1s together). A moment's thought will show that this means that the next line would have to be the same as the second line and so on, so that the output will be repeating the pattern up to the last line: i.e. the output will be 101111010000... repeating.

To get a 'feel' for the way that a scrambler works, you really need to look at the scrambling of various sequences for one or two different scramblers and for various different combinations of data in the storage elements at the start. If you have access to a computer, you could do this with a simple program. Appendix C suggests ways of using a spreadsheet.

Table 6.5. *Example scrambling by a 2-stage scrambler*

Starting state		Input data		
Stage 1	Stage 2	000000...	111111...	010101010101...
0	0	000000....	100100...	011100011100...
1	0	101101...	001001...	110001110001...
0	1	110110...	010010...	101010101010...
1	1	011011...	111111...	000111000111...

Table 6.5 gives the output patterns for the two-stage scrambler with each of the four possible starting states and for three different input sequences. For each of the input patterns, three of the possible starting states result in an output which is, to some extent, randomised. For example, for an all 0s input the starting states 10, 01 and 11 all result in outputs which are repeating patterns of length 3. But for each input there is one starting state for which the scrambler does not randomise the data. For the all 0s input this is the starting state 00, for which the output is all 0s. This effect of the scrambler failing to scramble is known as scrambler lock-up, or latching. Although most dramatic for all 0s or all 1s input, it also occurs for any repetitive input. From Table 6.5, for example, you can see that when the input is 0101 repeating the output is a repeating pattern of length 6, except for the case of the 01 starting state when the output is 1010 repeating – no better than the input.

The randomising effect of such a short scrambler as this is not very impressive, but it becomes much more useful as the scrambler length increases. In general, the (unlatched) output sequence of a suitably designed n-stage scrambler in response to an all 0s or all 1s input is a repeating sequence of period $2^n - 1$ bits. A seven-stage scrambler with taps from stage 6 and 7 for example, responds to an all 0s input with the 127-bit sequence:

0100000110000101000111100100010110011101010011111010000111 0 0010010110110101101111011000110100101110111001100101010111 1 111000001

The pattern of 1s and 0s in this sequence, furthermore, has mathematically defined randomness properties – such as the fact that the number of 1s differs from the number of 0s by no more than one. It is an example of the class of sequences known as *pseudo-random binary sequences* (*PRBS*). More about this aspect of such sequences will be found in Golomb.

As well as making the output more random, increasing the scrambler length also reduces the probability of lock-up. For any length of scrambler, and for every input pattern, there is just one starting state for which lock-up occurs. For the two-stage scrambler this meant that there were three starting states which result in a randomised output. For the seven-stage scrambler, there are 127 starting states which give a randomised output, compared to the one which causes lock-up. Often it can be assumed that the input data will be varying in such a way that the occurrence of lock-up can be considered to be a random statistical process, and by making the scrambler long enough the probability can be made small. In addition, other features can be designed into the scrambler which will ensure that lock-up will only cause problems for highly improbable data sequences (see Savage for a discussion of this).

6.3.2   Scrambler specification

Scramblers are usually described in, for example, CCITT recommendations by what is known as the *tap polynomial*.

For a two-stage scrambler of length n with taps from shift register stages m and n (i.e. the scrambler of Fig. 6.6) this would be of the form

$$1 + x^{-m} + x^{-n}$$

The feedback tap information is contained in the power of the terms included. The 1 is always present and represents the scrambler input. The polynomial for the two stage scrambler (Fig. 6.8) is therefore

$$1 + x^{-1} + x^{-2}$$

> **Differential coding is thus described by the polynomial $1 + x^{-1}$.**

This apparently rather odd way of describing a scrambler derives from an algebraic technique used for analysing systems that manipulate bit sequences by shift registers. The technique involves writing the bit sequences themselves as polynomials, with the 1s and 0s of the bit sequence taken as the coefficients of the terms of the polynomial. The 'x' in the polynomial can be interpreted as a shift operator, such that x^{-n} represents a delay of n bit intervals. Thus $1 + x^{-2}$ is interpreted as 1 'now', followed by 1 two intervals later; giving 101. The operation of the scrambler (starting with 0 on all stages) on any input sequence can then be determined by dividing the input bit sequence by the scrambler polynomial. The division uses modulo two arithmetic, in which addition is equivalent to the exclusive-OR logic operation, and multiplication is equivalent to the AND logic operation. For example, the start of the encoding shown in Table 6.4 is reproduced in polynomial form as

$$
\begin{array}{l}
\qquad\qquad 1+\qquad x^{-2}+x^{-3}+x^{-4}+x^{-5}+\qquad\qquad x^{-7} \\
\overline{1+x^{-1}+x^{-2}\,\big|\,1+x^{-1}+\qquad\qquad x^{-4}+x^{-5}+\qquad\qquad x^{-8}+x^{-9}+\dots} \\
\qquad\quad \underline{1+x^{-1}+x^{-2}} \\
\qquad\qquad\quad x^{-2}+\qquad x^{-4}+x^{-5}+\qquad\qquad x^{-8}+x^{-9}+\dots \\
\qquad\qquad\quad \underline{x^{-2}+x^{-3}+x^{-4}} \\
\qquad\qquad\qquad\quad x^{-3}+\qquad x^{-5}+\qquad\qquad x^{-8}+x^{-9}+\dots \\
\qquad\qquad\qquad\quad \underline{x^{-3}+x^{-4}+x^{-5}} \\
\qquad\qquad\qquad\qquad\quad x^{-4}+\qquad\qquad\qquad x^{-8}+x^{-9}+\dots \\
\qquad\qquad\qquad\qquad\quad \underline{x^{-4}+x^{-5}+x^{-6}} \\
\qquad\qquad\qquad\qquad\qquad x^{-5}+x^{-6}+\qquad x^{-8}+x^{-9}+\dots \\
\qquad\qquad\qquad\qquad\qquad \underline{x^{-5}+x^{-6}+x^{-7}} \\
\qquad\qquad\qquad\qquad\qquad\qquad x^{-7}+x^{-8}+x^{-9}+\dots \\
\qquad\qquad\qquad\qquad\qquad\qquad \underline{x^{-7}+x^{-8}+x^{-9}} \\
\qquad\qquad\qquad\qquad\qquad\qquad\qquad\qquad \dots
\end{array}
$$

The result is $1+x^{-2}+x^{-3}+x^{-4}+x^{-5}+x^{-7}+\dots$, which represents the sequence 10111101..., as required.

EXERCISE 6.2

Derive the output from a 3-stage scrambler with the tap polynomial $1+x^{-1}+x^{-3}$ when the input is the repeating sequence 1010...:

(a) if the scrambler starts with a 0 in the first and third stages, and a 1 in the second;

(b) if the scrambler starts with a 1 in the first stage and a 0 in the other two.

What has happened in the first case? (You could either make use of a computer program to derive the output, or else use a table like Table 6.4. Polynomial division is not appropriate because the starting state does not have 0 in all stages.)

6.3.3 Synchronisation

As with differential coding, because the code symbols are fed into the storage elements at both the scrambler and descrambler shift registers, scrambling is self-synchronising. For an n-stage scrambler, n symbols need to have been fed into the shift registers in order to synchronise them fully, so n bits need to be transmitted before synchronisation has been effected.

6.3.4   Error extension

For any scrambler with two taps, a single error in the coded data will result in three errors in the output. Hence scrambling has an error extension factor of 3.

This can be seen by considering the effect of a single error arriving at the descrambler. On first arriving at the descrambler, an erroneous symbol provides a wrong input to the exclusive-OR for the output, causing an error. It then enters the shift register and moves along stage by stage. On passing each tap, it causes an error into the feedback which causes another error on the output. If there are two taps, there will be a total of three errors.

6.3.5   'Set–Reset' scramblers

Set–reset scrambling is non-redundant, so all 0s or all 1s is still a valid output code sequence. However, it will only occur if the input data is a PRBS sequence of the correct type and with the correct alignment. This is a highly improbable occurrence for 'real' data, but could occur in the network because of the widespread use of PRBS sequences for testing – and scrambling in other parts of the system!

An alternative method of scrambling, sometimes called 'Set–Reset' scrambling, removes the error extension of self-synchronising scrambling, at the cost of requiring additional synchronisation circuits. Set–reset scrambling also, perhaps more importantly, does not latch for all 0s or all 1s input.

The randomising method used in set–reset scrambling is that of combining the input data with a known 'random' (pseudo-random) sequence in an exclusive-OR gate. Thus a 'random' selection of bits in the input data are inverted (wherever a 1 appears in the pseudo-random sequence), transferring the randomness of the pseudo-random sequence to the data (Fig. 6.9). The original data is recovered at the receiver by combining (in an exclusive-OR gate) the coded data with an identical replica of the pseudo-random sequence used at the transmitter – thus re-inverting those bits which were inverted at the transmitter.

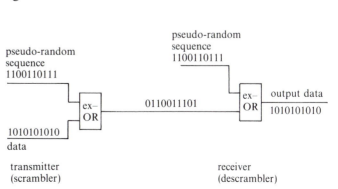

Fig. 6.9. Set–reset scrambling.

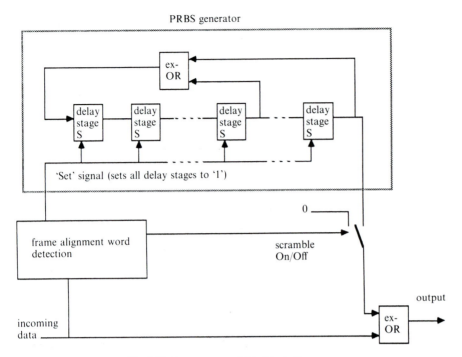

Fig. 6.10. Set–reset scrambling block diagram.

Fig. 6.10 shows the design of a set–reset scrambler in more detail. The PRBS generator is equivalent to a self-synchronising scrambler with continuous all 0s input (see Section 6.3.1).

The PRBS sequences must be aligned at the scrambler and descrambler, hence the need for synchronisation. This is usually achieved by making use of some known characteristic of the conveyed data. For example, if the conveyed data is a multiplexed signal, the frame alignment word can be used as a reference location in the bit stream. This is the technique used for the system shown in Fig. 6.10. At the end of the frame alignment word, the PRBs generators at both scrambler and descrambler are set to all 1s, then left to run continuously until the next frame alignment word. Since they start off in the same state at the same point in the bit stream, the two generators are aligned. In order for the descrambler to be able to find the frame alignment word, this section of the data is left unscrambled, hence the 'scrambling on/off' in Fig. 6.10.

Multiplexing and frame alignment words are discussed in Chapter 7. The significance here is just that a known pattern of bits occurs at intervals in the bit stream.

6.4   Block codes: error detection and error correction

As with line block codes, channel block codes work by taking a block of k input bits and encoding them to n output bits. Such a code is described as an (n,k) code and has efficiency $k/n \times 100\%$. Frequently, however, a channel code efficiency is expressed not as a percentage but as a fraction $\frac{k}{n}$; referred to as the *code rate*. The redundancy of a channel block code can also be specified; as with line code it is given by $(1 - \text{efficiency})$.

The idea that you can detect and even correct errors in digital systems can seem rather surprising when first encountered. The principle, however, is very simple. Error detection can be most easily performed by *parity* checks. A single parity bit added to a block of data can allow the receiver to detect when a single error has occurred in the block. With even parity, for example, the parity bit added will be 1 if there are an odd number of 1s in the block of data, and 0 if there are an even number of 1s. Thus the total block (data plus parity bit) will always have an even number of 1s. If a single error occurs, turning either a 1 to a 0 or a 0 to a 1, then the block will have an odd number of 1s and the receiver will know that this is wrong and that an error must have occurred. Of course, if two errors occur (or any even number of errors) then the parity will be correct and the receiver will not know about the errors. There are many codes which have been devised to improve upon the error detection capabilities of parity checks. One very commonly used class of error detection codes, cyclic redundancy checks, is discussed below.

Error detecting codes are often used in systems in which the receiver can communicate back to the sender. If there have been errors in the received signal, the receiver automatically alerts the sender so that the data can be retransmitted. (A technique known as ARQ: Automatic Repeat reQuest). In some systems ARQ is inappropriate (no backward path exists, or else the delay introduced by having to request retransmission is unacceptable, for example), so *(forward) error correction* is used.

Error correction can be illustrated by an extension of the use of parity checks. Fig. 6.11 shows a block of data arranged in a rectangle, with parity bits added for the parity of each row and each column, even parity being used in each case. The bottom right hand bit can be worked out using the

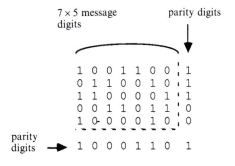

Fig. 6.11. Rectangular block code.

last row or the last column. Either way should give the same result because it is, in effect, a parity check over the whole block. Any single error in the data will show up as a parity error in one column and one row. The receiver can use the row and column locations as coordinates to find the offending bit and correct it. This kind of coding, which is known as *rectangular*, can correct any single error in a block, but can lead to problems if more than one error occurs in a block. For multiple errors it may be possible to detect the presence of errors without being able to locate and correct them, but in some cases groups of errors can completely disappear. No error correction or detection scheme can protect against *all* possible combinations of errors, but many schemes which are better than rectangular coding have been devised. Two classes are discussed in this chapter: Hamming codes and convolutional codes.

6.4.1   Cyclic redundancy checks

The principle behind CRCs is very simple. The data to be transmitted is viewed as single number. This number is divided by a pre-assigned constant (the *generator*, *G*, of the code), then the remainder (*R*) resulting from this division is recorded. When the data is transmitted *R* is also sent. The receiver performs the same division by *G*, and calculates a value for the remainder. If the value calculated at the receiver does not agree with the received value of *R*, then there must have been one or more errors.

The elegance of the system comes from a number of factors. Firstly, judicious choice of *G* leads to a very efficient code: that is, one which can detect large numbers of errors with the minimum necessary redundancy. Secondly, when applied as described below, the scheme can be implemented very neatly with simple electronic logic circuits. It is this latter point which has led to the widespread application of CRCs in digital communication links.

The implementation in digital systems works on the principle outlined above but uses modulo two arithmetic. The generator, the input data and the remainder are all expressed as modulo two polynomials, using the ideas described in Section 6.3.2. Working in modulo two, the CRC principle is particularly neat.

Consider now the Euclidean division algorithm:

$$\text{dividend} = (\text{quotient}) \cdot (\text{divisor}) + \text{remainder}$$

Rearranging gives

$$\text{dividend} - \text{remainder} = (\text{quotient}) \cdot (\text{divisor})$$

This algorithm merely formalises the common understanding of division. For example, 13 divided by 5 gives 2 with a remainder of 3. 13 is the dividend, 2 the quotient and 5 the divisor, so the Euclidean division algorithm states:
$13 = 2 \times 5 + 3$.

In modulo two, subtraction and addition give the same result, so we can write:

$$\text{dividend} + \text{remainder} = (\text{quotient}) \cdot (\text{divisor})$$

The significance of this formula is that if we add the dividend and the remainder we get a number ((quotient)·(divisor)) which is divisible by the divisor *with no remainder*.

In the system for CRCs, the quotient is G, the dividend is the input data (which we will call M) and the remainder is R: both M and R are transmitted to the receiver. So, instead of the receiver dividing M by G and comparing the resulting remainder with R, it adds M and R, divides by G, and just has to check to see if the result is 0. If it is not 0, then there must have been one or more errors.

There is one further step which simplifies the digital system. Suppose that the generator (G) has $r+1$ bits, with the most significant bit equal to 1. (Otherwise it would have fewer than $r+1$ bits.) The remainder, when it is formed, has at most r bits, that is one fewer than G. If the remainder has fewer than r bits, leading 0s are added to make the number of bits up to r.

Since we know how many bits there are in the remainder we can use the following ploy. Instead of performing the CRC coding and decoding on the input data M directly, we first create a new number, M', by shifting M by r places to the left. The 'new' r least significant bits of M' are all set to 0. We then calculate R (by dividing M' by G), and immediately add R to M': resulting in a number in which the r least significant bits are the remainder while the remaining bits are the original data word M. The number $M' + R$ is transmitted. The receiver divides the number it receives by G and if the remainder is 0 it knows that there have been no errors so it cuts out the r least significant bits and is left with the required data. If the remainder is not 0, then there must have been one or more errors and re-transmission will be requested (assuming the error detection is being used as part of an ARQ error control scheme).

The overall process is summarised in Fig. 6.12. Fig. 6.13 shows the implementation of a CRC using shift registers and exclusive OR gates. As explained in Section 6.3.2, a scrambler can be used to perform modulo two division. However, the circuit needed here has to be slightly different from the scramblers previously described, because the quantity wanted is the remainder following division, not the quotient. Conveniently, a circuit of the form shown in Fig. 6.13 performs the required division, *and* finishes up with the remainder as the pattern of bits left in the storage stages. Notice the difference between this circuit and the scramblers of Section 6.3.2; the 'taps' are here points where the feedback feeds *into* the shift register.

The circuit of Figure 6.13 is described by the polynomial $x^3 + x + 1$. This is equivalent to $(1 + x^{-2} + x^{-3})x^3$, and, indeed, the (quotient) output from this circuit for any input is the same as the output from a scrambler with the tap polynomial $1 + x^{-2} + x^{-3}$. The multiplying factor of x^3 affects the interpretation of the polynomial algebra, but does not alter the output bit sequence.

Transmitter

M 101011011
M' 101011011000

calculate R: $\dfrac{101011011000}{1011}$ =   100110001   +   011

                                           quotient   remainder
                                          (not needed) (R)

$M'+R$ = 101011011011

Receiver

received 101011011011

divide by G $\dfrac{101011011011}{1011}$ =   100110001   +   000

                                           quotient   remainder
                                          (not needed) $(0 \Rightarrow$ no errors)

output data 101011011

Fig. 6.12. CRC with generator $x^3 + x + 1$. (Note: when these words are transmitted the most significant bit, i.e. the left-most bit, is sent first.)

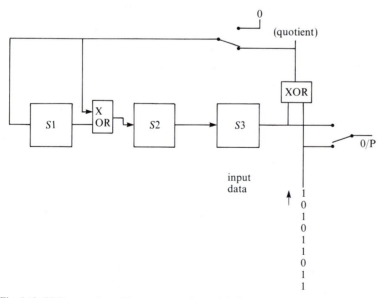

Fig. 6.13. CRC generation with generator polynomial $x^3 + x + 1$. Once all the data has been fed in the two switches change over so that the remainder is shifted out from the storage elements.

The procedure therefore is to feed the input data into the 'scrambler' while simultaneously transmitting the input data to the transmission line. At the end of the input data the switches are changed, as indicated in Fig. 6.13, so that the remainder is shifted out of the scrambler to the line. The receiver has a duplicate of the scrambler which divides the received data by the generator. At the end of the receiver's division, error-free data will be indicated by all storage elements in the scrambler being zero.

The defining feature of any particular cyclic redundancy check code is the generator, and it is usual to express this as the tap polynomial of the scrambler used for the division; this is the *generator polynomial* of the code.

Appendix C shows how simple CRC checks can be easily simulated on a spreadsheet.

6.4.2   Hamming codes

As stated at the beginning of this section rectangular codes are not very efficient as error correction codes. Consider the (48,35) code of Fig. 6.11. It can only be guaranteed to correct a single error in the block. Thus it uses 13 parity check bits to locate one of 48 possible errors. Now since each parity check can 'pass' or 'fail', there are $2^{13} = 8192$ different possible outcomes for the 13 parity checks, and only 48 of these outcomes are used to correct errors, although many more can be used to detect errors.

Consider now a code that uses m parity checks. There are 2^m different possible results for the parity checks. Assuming that there is one error at most, so that the case of no parity failures corresponds to there being no error, there remain $2^m - 1$ combinations, each of which could perhaps be used to identify a different error. This would lead to a very efficient code but it would only be possible provided that the total number of bits, n, including the parity bits, did not exceed $2^m - 1$, which gives the condition

$$2^m \geq n + 1$$

There exist codes which meet the equality condition: such codes are special cases of what are known as *perfect codes*.

This is the definition of 'perfection' for single error correcting codes. A more general definition covers 'perfection' in codes designed to correct more than one error in a block (see, for example, Clark and Cain). Perfect codes are not necessarily always the best codes to use in any given application. It depends upon exactly what features are required from the code.

Hamming codes are one of the few cases of perfect codes. The idea is to have several parity checks, each relating to a particular grouping of some of the message bits. Assume that the various checks are carried out when the coded message is received and that the results are expressed as a 0 if the test is successful and 1 if it fails. Imagine that the results of the tests are written as a string of these 0s and 1s to form, in effect, a binary number. If the checks are suitably arranged, this number – the *syndrome* – can directly indicate the position of the erroneous bit in the received message. Thus with 3 parity bits, we can have $2^3 - 1 = 7$ bits in all and, if the

Table 6.6. *Hamming codes for the 16 4-digit message words*

Original message				Coded message 1	2	3	4	5	6	7
0	0	0	0	0	0	0	0	0	0	0
0	0	0	1	1	1	0	1	0	0	1
0	0	1	0	0	1	0	1	0	1	0
0	0	1	1	1	0	0	0	0	1	1
0	1	0	0	1	0	0	1	1	0	0
0	1	0	1	0	1	0	0	1	0	1
0	1	1	0	1	1	0	0	1	1	0
0	1	1	1	0	0	0	1	1	1	1
1	0	0	0	1	1	1	0	0	0	0
1	0	0	1	0	0	1	1	0	0	1
1	0	1	0	1	0	1	1	0	1	0
1	0	1	1	0	1	1	0	0	1	1
1	1	0	0	0	1	1	1	1	0	0
1	1	0	1	1	0	1	0	1	0	1
1	1	1	0	0	0	1	0	1	1	0
1	1	1	1	1	1	1	1	1	1	1

syndrome comes to, say, 101, then the bit number 5 was received in error. If the syndrome is 000, all parity checks succeeded and there was no error. Such a code would have $7-3=4$ data bits and the construction is as follows.

The data bits are put in positions 3, 5, 6 and 7; the parity bits in positions 1, 2 and 4. The parity checks are as follows:

check 1 is bit 1 and gives even parity to bits 1   3   5   7
check 2 is bit 2 and gives even parity to bits   2 3   6 7
check 3 is bit 4 and gives even parity to bits   4 5 6 7

The syndrome is then: check 3; check 2; check 1. So if bit 1 is in error, only check 1 fails and the syndrome is 0 0 1, pointing to check bit 1. If bit 3 is in error, both check 1 and 2 fail, giving the syndrome 0 1 1, pointing to bit 3. You can verify that any other single bit in error will generate the correct syndrome. The resultant code table is shown in Table 6.6.

EXERCISE 6.3

Decode the following received words, which have been encoded by the (7,3) Hamming code just described (assume that they are either error free or contain one error).

(a) 1110100
(b) 0011100
(c) 0010110

There is an algorithm for constructing Hamming codes for any number of parity bits, such that the syndrome operates as described above. This is quite straightforward but will not be described here: it can be found in any of the standard texts (see, for example The Open University (1990)).

The Hamming code will still work if a different order is used for the message and parity bits, but the position of the error will no longer be given directly by the syndrome. This does not matter if a look-up table is used and if, for instance, the encoder treats the 4 data bits as an address in which the corresponding 7 bit Hamming code is stored. Similarly, the decoder can use a look-up table with the received 7 bit words as addresses for the correct 4 bit source messages.

> **This is a similar method of coding and decoding to that used by line block codes (section 5.2.4).**

Hamming distance Taking the 7 bit Hamming code as an example, there are 4 data bits so that $2^4 = 16$ different messages can be sent. These 16 messages, and the correct Hamming codes for them are shown in Table 6.6. With 7 bits in the Hamming coded words, there are $2^7 = 128$ combinations in all, of which only 16 represent correctly coded words. It is this high degree of redundancy which makes error correction possible. If you study Table 6.6 you will see that every correctly coded word differs from every other one by at least 3 bits. The significance of this minimum difference of 3 bits is that, if there is only one error, the received pattern will differ by only 1 bit from the correct transmitted pattern, but it will differ by at least 2 bits from all the others. The correct transmitted word can therefore be deduced. In the theory of error correcting codes, the term Hamming distance is used for the number of digits by which any two correct code words differ. If the Hamming distance between any two correct codes is 5, then, if two errors occur, the received message will differ from the correct one in 2 places, but from all others by at least 3 places. A code with a Hamming distance of 5 can therefore be used to correct 2 errors. Similarly a code with a Hamming distance of 7 can correct 3 errors.

6.5   Convolutional coding

Convolutional coding is another channel coding technique for error correction. It differs fundamentally from cyclic redundancy checks by not being a block code. As with scrambling, it operates on the input data continuously, so that the encoding of any input bits depends upon the encoding that has gone before: it is said to have *memory*.

The name 'convolutional' is used because the code generation is essentially a type of discrete convolution (Chapter 2) using modulo arithmetic.

Convolutional codes are often constructed with high redundancy to give good error correction. They generally allow easy construction of encoders and decoders, and are used where high rates of error correction are required from a minimum of circuitry; for example, in satellite systems.

Fig. 6.14 shows a simple convolutional encoder. Before going any further, look at the difference between this and a scrambler (Fig. 6.6). In addition to there being more than one output, a significant difference is that there is no feedback in the shift register. The sequence entered into the shift register is the input sequence, not the output as for a scrambler.

Notice the three-input exclusive-OR gate used by the convolutional coder. Multiple input exclusive-OR gates are common in many digital communication systems. The truth table for all exclusive-OR gates can be summarised by saying that the output is 0 if an even number of inputs are logic 1 and is 1 if an odd number of inputs are logic 1.

CRCs and Hamming codes are memoryless coding techniques. However, some of the line block codes (such as 3B4B) do have memory because the choice of code word depends upon the current disparity state – which is affected by what coding has gone before.

6.5.1 Encoding

The data is entered into the shift register one bit at a time. After entering in each single bit, three output symbols are read in succession: A, B, C in Fig.

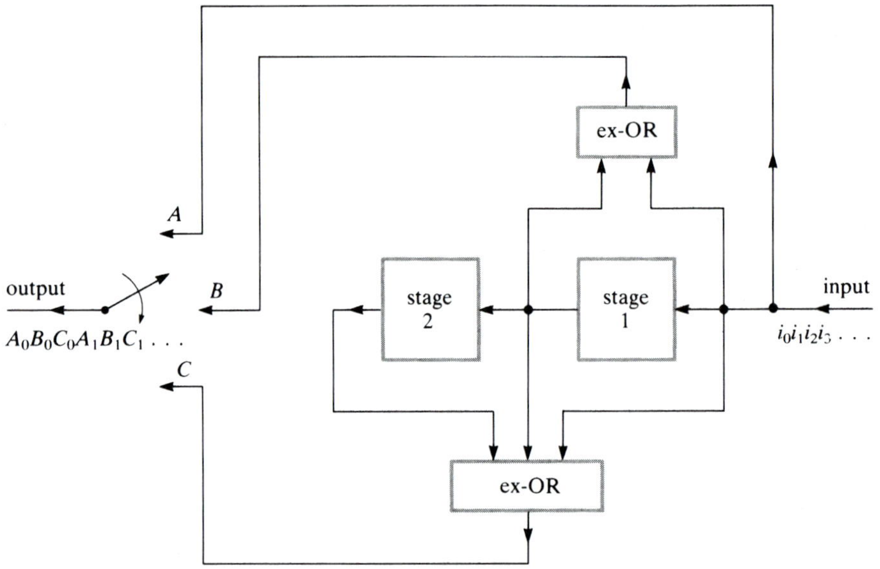

Fig. 6.14. A convolutional coder (note that this has been drawn so the data enters from the right).

Table 6.7. *The coding of 100111*

Input bit	Shift register contents 1 2		Output symbols *A B C*	
1	0 0		1 1 1	
0	1 0		0 1 1	
0	0 1		0 0 1	
1	0 0		1 1 1	
1	1 0		1 0 0	
1	1 1		1 0 1	

So the serial output data are: 111011001111100101

6.14. In this code, *A* is the same as the input bit. As an example, the encoding of 100111 is shown in Table 6.7. Since this code produces three binary coded symbols for each input bit, it has an efficiency of 33%. The number of input bits used in the encoding of a set of channel symbols, that is the length of the shift register plus 1 for the new input bit, is called the constraint length of the code.

When a new input bit arrives, the shift register already contains the previous two input bits in stages 1 and 2. The output corresponding to each new input bit therefore depends upon this input bit, and the two previous input bits. The shift register may be thought of as existing in one of four possible states, corresponding to the four possible combinations of the bits in stages 1 and 2. The output of the encoder will then depend upon this state, and on the value (1 or 0) of the newly arrived bit. All possible outputs can be tabulated as in Table 6.8.

When the three output symbols resulting from a new input bit have been transmitted, the input bit will be shifted into stage 1 of the shift register, and the contents of stage 1 will be shifted into stage 2. In other words the state of the encoder will change, and the new state will be determined by the value of the new bit.

From Table 6.8 it is possible to deduce what the new state will be for each input bit, and this has been shown by extending Table 6.8 to give Table 6.9. Although there are eight entries in the 'next state' columns of Table 6.9, there are, of course, still only four states in which the encoder may exist. In fact each of these four states appears twice in the 'next state', column because there are two ways each can be reached.

The contents of Table 6.9 can be displayed graphically, drawing a line between a point representing the current state and a point representing the next state, for each of the four possible current states and both values

Table 6.8. *The coding of a convolutional encoder*

Encoder state			Output
Stage 2	Stage 1	New input	A B C
0	0	0	0 0 0
		1	1 1 1
0	1	0	0 1 1
		1	1 0 0
1	0	0	0 0 1
		1	1 1 0
1	1	0	0 1 0
		1	1 0 1

Table 6.9. *The convolutional encoder coding extended to show the new state of the encoder after encoding an input bit*

Current state			Output	Next state	
Stage 2	Stage 1	New input	A B C	Stage 2	Stage 1
0	0	0	0 0 0	0	0
		1	1 1 1	0	1
0	1	0	0 1 1	1	0
		1	1 0 0	1	1
1	0	0	0 0 1	0	0
		1	1 1 0	0	1
1	1	0	0 1 0	1	0
		1	1 0 1	1	1

of the next bit. This has been done in Fig. 6.15. From each current state there are two lines, one (drawn as the upper line from each current state) caused by the input being a 0, and one (drawn as the lower line from each current state) caused by the input being 1. The output symbols are shown above the line connecting the two states.

Of course, the new state is the current state as far as the next bit that arrives is concerned, so Fig. 6.15 can be extended for subsequent bits, to arrive at a graph like Fig. 6.16. For obvious reasons, this is known as a trellis diagram.

State
description

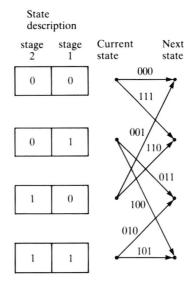

Fig. 6.15. Graphical representation of Table 6.9.

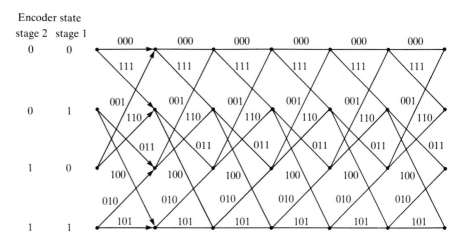

Fig. 6.16. A trellis diagram for the coder of Fig. 6.14.

Fig. 6.16 shows all possible encodings for any input data. The coding for particular input data can be found by following the appropriate path through the trellis. Starting at some initial state, the lower of the two possible paths is taken each time a 1 is at the input and the upper of the two possible paths is taken each time a 0 is at the input. The output data for each input bit is then given by the three symbols above the relevant transition.

EXAMPLE

What is the encoding of 100111 by this convolutional code (assuming that the coder starts with 0 in both stages)?

Solution. The path can be followed through on the trellis, starting in the top left hand corner. This has been done with the bold line in Fig. 6.17, and reading the output data from above the path gives the output: 111 011 001 111 100 101.

An alternative representation equivalent to the trellis is a state diagram. This can be constructed from Fig. 6.15 by superimposing the column of 'next states' on the column of 'current states' so that the lines showing the transitions now loop round to the appropriate state. This is shown in Fig. 6.18, where the transitions due to input 1s are indicated by dashed lines and transitions due to 0s by continuous lines.

EXERCISE 6.4

Check that the state diagram (Fig. 6.18) leads to the same output sequence as the trellis (Fig. 6.17), for the input sequence 100111.

6.5.2 Decoding and error correction

There are several different possible approaches to the decoding of data encoded by convolutional coding. In the literature on the subject you will find them described as sequential decoding, threshold decoding and Viterbi decoding. This section will look in some detail at Viterbi decoding since it is one of the most commonly used techniques in modern systems. It is named after A.J. Viterbi who first applied the technique for decoding convolutional codes in 1967.

For any decoding, the decoder first has to synchronise to the symbol blocks. In the example of Fig. 6.14, this means identifying the groups of three output symbols so that the decoder knows which is which of *A*, *B* & *C*. The following discussion assumes that this has already been done.

In essence, error detection and correction are based upon detecting coding rule violations. As with error detection through measuring the running digital sum in line codes (Chapter 5), the decoder follows the encoder so that in the absence of errors the decoder will always be in the same state as the encoder: any discrepancy implies an error. The decoder can be thought of as plotting the path of the received data on a copy of the code trellis. If there are no errors, the path through the trellis taken at the

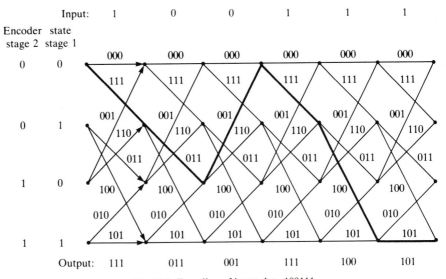

Fig. 6.17. Encoding of input data 100111.

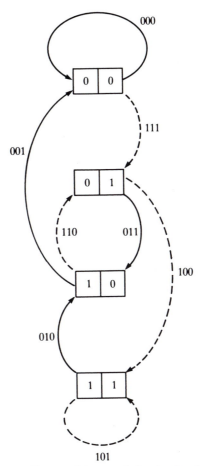

Fig. 6.18. State diagram of the convolutional coder of Fig. 6.14.

receiver will be the same as the path taken at the transmitter. An error will change the path at the receiver, and will sooner or later result in the receiver getting data which implies an impossible path. The receiver can then flag an error, and may be able to deduce which path is most likely to have been intended, thereby correcting the error. The decoder chooses the most likely path on the 'minimum distance' criterion: the data that implies the fewest line errors (minimum Hamming distance between intended and received data) is most likely to be the intended data.

In order to explain the principle of Viterbi decoding, the following discussion looks at decoding of data which has been encoded by the simplest possible convolutional coder: that shown in Fig. 6.19. This provides two output symbols for each input bit, and has just two states, determined by whether the bit in the storage element is a 0 or a 1. Its trellis diagram is shown in Fig. 6.20. The data which has been received is the sequence 11 11 11 10 10 01 and the path of this data has been plotted by a bold line on Fig. 6.20. Also shown on this figure are the time intervals, t_1 to t_6, identifying the arrival of each pair of symbols, and the output data. The values given for the output data are those given by taking the received data at 'face value', i.e. plotting the appropriate transition and decoding to a 0 if the transition is an upper line, to a 1 if it is a lower line.

The data which had been intended (that is, the data which was transmitted) is shown in Fig. 6.21. Comparing Fig. 6.21 with Fig. 6.20 you can see that there has been an error in the third symbol, and this has led to the decoded output data at time t_2 being wrong (1 instead of 0). Viterbi decoding leads to the decoder giving the output of the path of Fig. 6.21 when it receives the data of Fig. 6.20. It is possible, of course, that when the data of Fig. 6.20 is received, it was not intended to be that of Fig. 6.21. But in that case there would have to have been more than just one error in the code sequence. On the minimum distance criterion therefore (see Chapter 5), the data of Fig. 6.21 should be given as the output.

It is clear to the receiver at time t_2 that there has been some error in the encoded data because the receiver is in state 1 and is therefore expecting

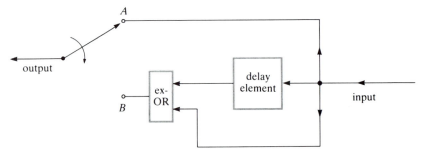

Fig. 6.19. A very simple convolutional coder.

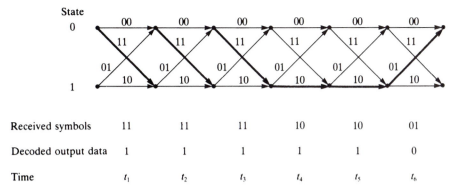

Received symbols	11	11	11	10	10	01
Decoded output data	1	1	1	1	1	0
Time	t_1	t_2	t_3	t_4	t_5	t_6

Fig. 6.20. Received data plotted on a trellis.

either 01 or 10, but it actually receives 11. A Viterbi decoder therefore looks at what would happen if it assumes each of the possibilities it was expecting: 01 and 10.

The result of these assumptions is shown in Fig. 6.22, with the two possible paths labelled A and B. Below the trellis, for each path, is shown what the intended code sequence it is assuming is, and how far that intended sequence is from the sequence that is actually received. So path A assumes that 01 was intended at time t_2, which differs from 11 in one place, giving a distance of 1. For both A and B, the total Hamming distance is 1, so there is no indication, at this stage, as to what is the best assumption, and the decoder continues looking at all possibilities.

At time t_3, the decoder now looks at four possibilities, because paths A and B can each go on to two different directions. The two paths following A, are labelled AA and AB, and the two following B, BA and BB, these are shown in Fig. 6.23. For each of these four paths the sequence they are assuming is again shown below the trellis together with the distance of this assumed sequence from the received sequence. So assuming path AA, the

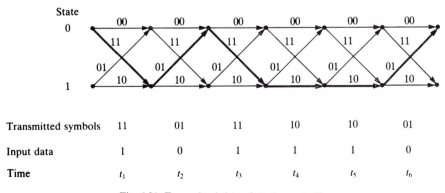

Transmitted symbols	11	01	11	10	10	01
Input data	1	0	1	1	1	0
Time	t_1	t_2	t_3	t_4	t_5	t_6

Fig. 6.21. Transmitted data plotted on a trellis.

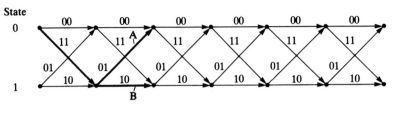

Received symbols	11	11
Path A: Intended code sequence assumed		01
Distance from sequence actually received		1
Path B: Intended code sequence assumed		10
Distance from sequence actually received		1
Time	t_1	t_2

Fig. 6.22. Data plotted to time t_2.

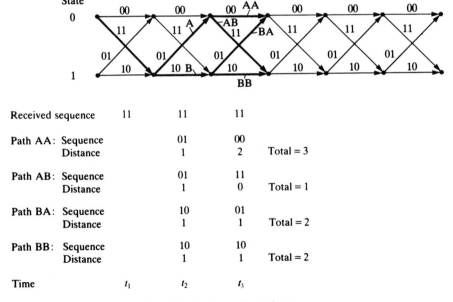

Received sequence	11	11	11	
Path AA: Sequence		01	00	
Distance		1	2	Total = 3
Path AB: Sequence		01	11	
Distance		1	0	Total = 1
Path BA: Sequence		10	01	
Distance		1	1	Total = 2
Path BB: Sequence		10	10	
Distance		1	1	Total = 2
Time	t_1	t_2	t_3	

Fig. 6.23. Possible paths at time t_3.

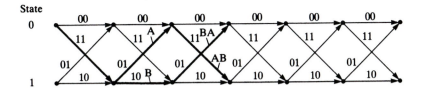

Received sequence	11	11	11	
Path BA: Sequence		10	01	
Distance		1	1	Total = 2
Path AB: Sequence		01	11	
Distance		1	0	Total = 1
Time	t_1	t_2	t_3	

Fig. 6.24. Survivor paths at time t_3.

distance is 1 at time t_2 and 2 at time t_3, giving a total distance for this path of 3. Although the paths now have different total distances, it is still too early to draw any final conclusions because further errors might appear subsequently for each path. However, some initial conclusions are possible.

Look at paths AA and paths BA. They both arrive at state 0 after time t_3. Since they are both now in the same state, whatever happens subsequently must be the same for each. But in getting to this state, path AA has incurred a distance of 3, while BA has incurred only 2. So whatever happens subsequently, path AA must always have a greater total distance than BA. Since the decoding criterion is to look for the path with the minimum distance path AA may be rejected.

Similarly, comparing paths AB and paths BB you can see that they both end up at state 1 but that path AB has the smaller total distance, so path BB is rejected. The two paths which have not been rejected (the *survivor* paths) are therefore BA and AB. These are shown in Fig. 6.24.

At time t_4 there are again four paths, because each of the two survivor paths from time t_3 has two possibilities at time t_4 (Fig. 6.25). The receiver again looks at the paths which end up at a common state in order to reject those with the largest distance. Paths BAB and ABB both end up at state 1, but BAB has the larger distance and is rejected. BAA and ABA both end up at state 0, but in this case they both have the same total distance of 3. Since they have the same distance, they have equal claim to being the right path and there is no way of choosing between the two – nor will there ever be, because future paths will not affect them up to this point. So in fact the choice between the two is arbitrary.

Suppose it is decided to reject BAA, then the two surviving paths are

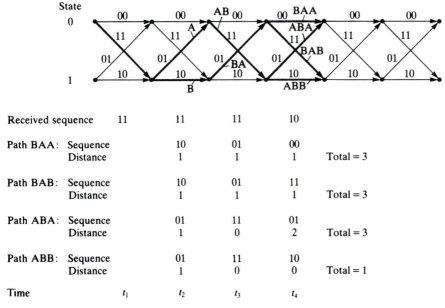

Received sequence	11	11	11	10	
Path BAA: Sequence		10	01	00	
Distance		1	1	1	Total = 3
Path BAB: Sequence		10	01	11	
Distance		1	1	1	Total = 3
Path ABA: Sequence		01	11	01	
Distance		1	0	2	Total = 3
Path ABB: Sequence		01	11	10	
Distance		1	0	0	Total = 1
Time	t_1	t_2	t_3	t_4	

Fig. 6.25. Possible paths at time t_4.

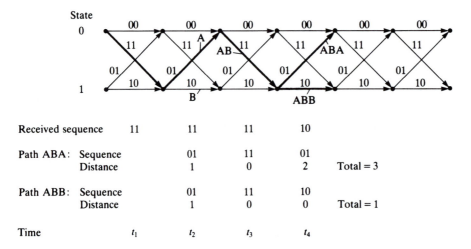

Received sequence	11	11	11	10	
Path ABA: Sequence		01	11	01	
Distance		1	0	2	Total = 3
Path ABB: Sequence		01	11	10	
Distance		1	0	0	Total = 1
Time	t_1	t_2	t_3	t_4	

Fig. 6.26. Survivor paths at time t_4.

ABA and ABB, as shown in Fig. 6.26. Looking at these two paths, you can see that they agree on the decoding at times t_2 and t_3. Furthermore, comparing Fig. 6.26 with Fig. 6.21, you can see that this agreed decoding is the correct one (that is, the best decoding based on the minimum distance criterion), so the error has been corrected. Having got an agreed decoding, the receiver need consider times t_2 and t_3 no longer: it provides the appropriate output and investigates the decoding from t_4 onwards.

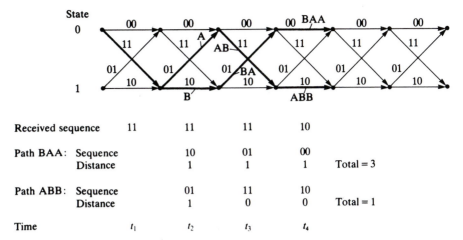

Received sequence	11	11	11	10	
Path BAA: Sequence		10	01	00	
Distance		1	1	1	Total = 3
Path ABB: Sequence		01	11	10	
Distance		1	0	0	Total = 1
Time	t_1	t_2	t_3	t_4	

Fig. 6.27. Alternative survivor paths at time t_4.

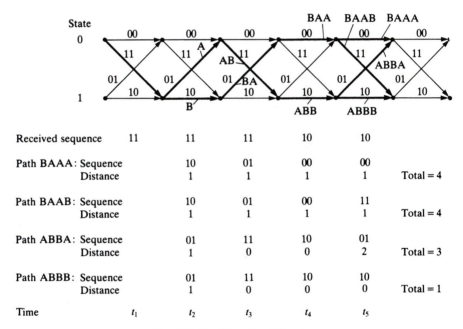

Received sequence	11	11	11	10	10	
Path BAAA: Sequence		10	01	00	00	
Distance		1	1	1	1	Total = 4
Path BAAB: Sequence		10	01	00	11	
Distance		1	1	1	1	Total = 4
Path ABBA: Sequence		01	11	10	01	
Distance		1	0	0	2	Total = 3
Path ABBB: Sequence		01	11	10	10	
Distance		1	0	0	0	Total = 1
Time	t_1	t_2	t_3	t_4	t_5	

Fig. 6.28. Possible paths at time t_5.

However, in going from Fig. 6.25 to Fig. 6.26, there was an arbitrary decision to keep path ABA in preference to BAA when they had the same distance: what would happen if BAA was kept instead?

This would leave the survivor paths BAA and ABB, as shown in Fig. 6.27. There is no agreed decoding in this case, so the decoder must continue to follow through the two paths from time t_2 onwards. The four paths at time t_5 are shown in Fig. 6.28.

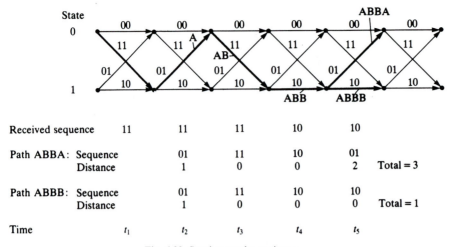

Received sequence	11	11	11	10	10	
Path ABBA: Sequence		01	11	10	01	
Distance		1	0	0	2	Total = 3
Path ABBB: Sequence		01	11	10	10	
Distance		1	0	0	0	Total = 1
Time	t_1	t_2	t_3	t_4	t_5	

Fig. 6.29. Survivor paths at time t_5.

Looking for survivor paths again results in the rejection of BAAA and BAAB, leaving ABBA and ABBB as shown in Fig. 6.29. There is now agreement over the decoding up to time t_5, and the error is corrected. So the arbitrary choice at time t_4 only affected how long it took to correct the error – it is corrected eventually whichever choice is made.

This decoding was started by suggesting that the decoder uses this method of trying out paths and looking for minimum distance when it knows there has been an error. In fact the decoder doesn't need to 'know' anything about the presence or absence of an error: it uses this decoding procedure constantly. In the absence of errors, the survivor paths rapidly agree on the correct decoding. To see this, look at the decoding for time t_6, as shown in Fig. 6.30. The error has already been corrected and the received data contains no more errors.

The survivor paths at this time are shown in Fig. 6.31, where you can see that they agree on the decoding for time t_5. Continuing in the absence of errors will always result in agreement over the decoding of the previous time. It is only in the event of an error that survivor paths maintain disagreement for longer. There are, however, still two paths for the current time. So if this were the end of a finite block of data, the last symbol would not have been decoded. This can be overcome by transmitting a predetermined bit at the end of the data, referred to as the tail digit. Knowledge of this last digit by the receiver, as well as removing uncertainty about the path at the last time interval, will ensure the decoding for earlier times if there are two paths still existing because of an earlier error.

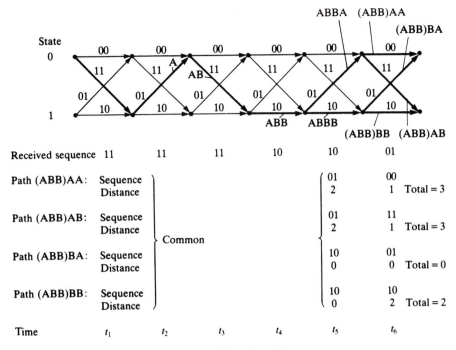

Fig. 6.30. Possible paths at time t_6.

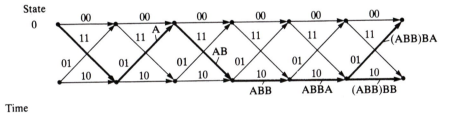

Fig. 6.31. Survivor paths at time t_6.

6.5.3 Practical convolutional codes

The two codes described in the previous section were examples of particularly simple convolutional codes. Although many such codes are used, more complex codes are also possible. In addition to codes using longer shift registers, codes where more than one bit is shifted into the coder at a time are possible. A general code which shifts in k bits and provides n code bits out would have efficiency $(k/n) \times 100\%$. Often, rather than referring to the efficiency of a convolutional code, the quantity k/n is often quoted as a ratio, and called the code *rate*.

diagram or a state diagram. The constraint length of a convolutional code is the number of input bits involved in the encoding of each block of output bits. Several different approaches to decoding are possible. Viterbi decoding is based upon following the received data on a trellis and observing when the received data implies an impossible path. Error correction may be possible by inferring the intended path, that is, by determining which path would result in the minimum distance between the symbols implied by the inferred path and the received path. Distance may be Hamming distance or an analogue measure of distance for soft decoding.

receiver. Differential coding, or a similar coding method, is also used prior to partial response signalling to minimise error extension.

Scrambling randomises data sequences, allowing highly structured data to be conveyed over channels which would otherwise degrade data containing repetitive patterns. Since, however, scrambling is non-redundant, randomness cannot be guaranteed unless there is a restriction on the possible input data sequences. In particular, self-synchronising scramblers can lock-up with an all 0s input if the scrambler shift register already contains all 0s. Set–reset scramblers cannot lock up in the same way as self-synchronising scramblers, but require some means of synchronisation between the transmitter and receiver.

Cyclic redundancy checks (CRCs), Hamming codes and convolutional codes all add redundancy for error control. CRCs provide error detection, but not correction. They are commonly used in systems where it is possible to retransmit data which has been found to contain errors at the receiver. Hamming codes and convolutional codes are error-correcting, and used in systems where forward error correction is required.

Both scramblers and CRCs can be analysed using polynomials. Self-synchronising scramblers continuously divide the data by the scrambler polynomial, transmitting the quotient one digit at a time. CRCs also divide the incoming data by a polynomial (the characteristic polynomial), but the transmitted data is the incoming data (the dividend) followed by the remainder from the division: the quotient is not used. In effect, error detection is performed by CRCs by recalculating the remainder at the receiver: if the remainder calculated at the receiver does not agree with the remainder calculated at the transmitter (and sent along with the dividend) then there must have been an error. For both scrambling and CRCs the necessary polynomial division is performed by shift registers with feedback taps, and the features of the code are determined by the choice of divisor polynomial.

Hamming codes are based upon interleaved parity checks. In the event of a single error, the pattern of checks which fail is the syndrome which points to the error. If two errors occur, the presence of errors will be detected, but they cannot be corrected. More than two errors may or may not be detected. For a general block error correcting code, the number of errors in a single block which can be corrected depends upon the Hamming distance between code words. To be able to correct n errors per block, the minimum distance between any two code words must be $2n + 1$.

Convolutional coding is, like scrambling, a continuous coding process (in contrast to block codes like CRCs and Hamming codes). However, convolutional coding adds redundancy, usually quantified by the code rate. The operation of a convolutional coder can be displayed on a trellis diagram or a state diagram. The constraint length of a convolutional code is the number of input bits involved in the encoding of each block of output bits. Several different approaches to decoding are possible. Viterbi decoding is based upon following the received data on a trellis and observing when the received data implies an impossible path. Error correction may be possible by inferring the intended path, that is, by determining which path would result in the minimum distance between the symbols implied by the inferred path and the received path. Distance may be Hamming distance or an analogue measure of distance for soft decoding.

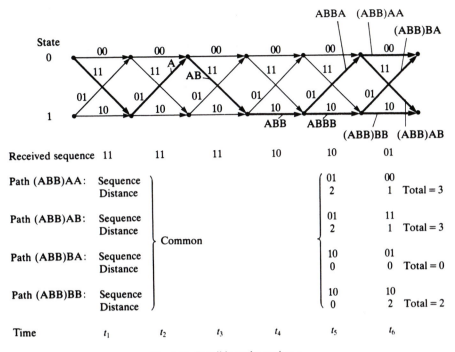

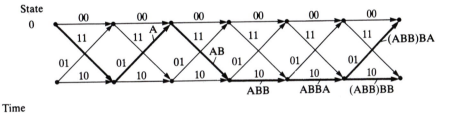

Fig. 6.30. Possible paths at time t_6.

Fig. 6.31. Survivor paths at time t_6.

6.5.3 Practical convolutional codes

The two codes described in the previous section were examples of particularly simple convolutional codes. Although many such codes are used, more complex codes are also possible. In addition to codes using longer shift registers, codes where more than one bit is shifted into the coder at a time are possible. A general code which shifts in k bits and provides n code bits out would have efficiency $(k/n) \times 100\%$. Often, rather than referring to the efficiency of a convolutional code, the quantity k/n is often quoted as a ratio, and called the code *rate*.

The description of Viterbi decoding given in the last section involved measuring the Hamming distance between possible paths and the received data. Instead of the Hamming distance, a 'soft distance' (see Chapter 5, Section 5.3.3) can be used to advantage. The decoding would be identical, but the distance would have more resolution than just a count of errors.

EXERCISE 6.5

Fig. 6.32 shows the trellis constructed by a receiver using Viterbi decoding when an error has occurred at time T_2. At time T_4 there are eight paths, which converge on to the four state. Calculate the distance of the eight paths from the received sequence (110 011 001, starting at time T_2) and draw a trellis showing the survivor paths.

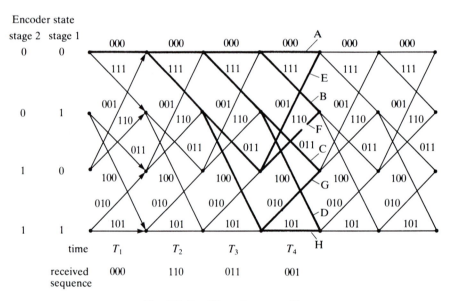

Fig. 6.32. Possible paths on a trellis.

6.6 Summary

The function of channel coding is to maintain the integrity of data transmitted over a communication channel.

Differential coding renders the data transparent to inversion. Combined with phase shift keying (PSK), it gives differential phase shift keying (DPSK) which removes the need for an absolute phase reference at a

Part 3

Digital transmission over the public switched telephone network

So far, this book has been concerned with models and processes which are relevant to a wide variety of telecommunication systems. The linear modelling tools of Part 1, and the digital techniques of coding, modulation and pulse processing discussed in Part 2, are applicable to line telephony, digital microwave links, satellite and mobile systems, and many other fields. In a book this size it is impossible to deal with all areas of modern digital telecommunications. To set the previous material into context, we have chosen therefore to concentrate in this final part on transmission aspects of just one, large-scale, system: the digital public switched telephone system or *PSTN*. The wide variety of topics with which a modern telecommunications engineer needs to be conversant is reflected particularly here, ranging from the borders of electronics almost to those of software engineering.

Fig. 1.1 of the Introduction showed part of an integrated services digital network (ISDN), in which signals from a variety of sources are transmitted over a universal network. Some of these signals are inherently digital (such as computer data), others are by nature analogue (such as speech input to a telephone handset). In an ISDN they are all transmitted as digital signals with a common format, and their origin is immaterial as far as network management is concerned.

There are currently (1992) few examples of true ISDNs. Nevertheless, in many countries the public switched telephone network is evolving in this direction. Fig. 3 shows part of such a telephone network. The boxes represent exchanges interconnected by numerous *trunk routes*. At the lowest level are local switching centres (local exchanges), to which individual user telephones, or private branch exchanges (PBXs), are connected via *local lines* (not shown in the figure). Above the local exchanges is a hierarchy of intermediate and international switching centres. Calls between individual users are interconnected by means of complex switching operations through intermediate exchanges, the precise switching centres involved being determined by both the geographical locations of the callers and the prevailing telephone traffic conditions. Telephone networks of this type have existed for many years, evolving gradually as analogue telephony became increasingly available to subscribers. Over the last decade or so, however, the trunk routes of

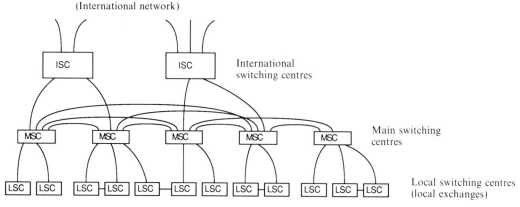

Fig. 3. Part of a public switched telephone network (PSTN).

PSTNs in many countries have been converted largely to digital transmission. A major task of the network providers has therefore become that of ensuring the transmission of bit streams from one node to any other at a suitably low error rate.

Currently, even in countries where trunk routes are almost entirely digital (optical fibre or digital microwave), few users have direct digital access to the PSTN, and most have to rely on local lines originally installed for analogue telephony (usually a pair of twisted copper wires). Telephone signals are therefore normally transmitted to the local exchange in analogue form, and converted there to a digital signal for onward transmission. When an inherently *digital* device such as a facsimile machine uses the PSTN, a modem (modulator/demodulator) first codes the digital information into sinusoidal pulses in the way described in Chapter 4. These sinusoidal pulses are then treated by the network exactly as though they were *analogue* speech signals – that is, they are themselves sampled at a higher rate, transmitted as a digital signal over a digital channel, reconstructed as sinusoidal pulses, and finally interpreted as digital data by the receiving device! Only by the eventual provision of a suitable digital link to end users can such anomalies be eliminated.

PSTN design and operation is extremely complex, and cannot be covered in detail here. The final three chapters of this book will therefore examine three particularly important aspects of digital transmission over the PSTN, drawing on the fundamental material of earlier parts;

> 1 The highest level is the interfacing of the signal source with the network. Chapter 7 looks at several functions of the PSTN at this level. The main description is of *pulse code modulation* and related

The need to rely (at least initially) on existing subscriber lines is a severe constraint on the introduction of ISDN services to end users. Various techniques are under investigation, or currently being introduced, to enable such twisted pairs to carry digital signals with a much higher bandwidth than the 4 kHz for which they were originally designed.

techniques for digital coding of analogue message signals. Also discussed are the uses of *modems* where the source is digital but has to be transmitted over analogue lines to the exchange, and interface standards for ISDN.

2 Once all signals are in digital form, from whatever source, they have to be transported efficiently over the whole network. At this intermediate level, the precise details of the links between nodes are unimportant; the task is to utilise bandwidth to the full and to ensure that the digital information remains synchronised. Chapter 8 considers this problem, and in particular how the emerging standard known as the *synchronous digital hierarchy* (*SDH*) copes with the task.

3 At the lowest level, digital pulses must be transmitted along a single link in a way which minimises error. Chapter 9 examines some features of the design of an optical fibre link of the type widely used in digital PSTNs, concentrating on how signal to noise ratio can be maximised (and thus error rate minimised) at the optical receiver.

These three aspects can be viewed as a hierarchy of functions – rather like the OSI model of the introduction and the layer model used in Part 2. As with those layer models, additional overheads are incurred as signals pass down the hierarchy. At the intermediate (SDH) level, various additional bits have to be incorporated into the bit stream to carry out alignment and synchronisation tasks; at the level of an individual link, line coding and perhaps error detection/correction overheads will be incurred. It should be useful to bear this in mind when reading Part 3, although there is no direct correspondence between the topics of the next three chapters and specific OSI layers.

7

Analogue and digital sources

7.1 Introduction

This chapter is concerned with the initial processing of analogue and digital signal sources prior to transmission over a public switched telephone network (PSTN). A major topic is *pulse code modulation* (*PCM*), a technique for encoding analogue signals and transmitting them over a digital link. (The name was originally chosen by analogy with other pulse modulation techniques such as pulse amplitude modulation (PAM) and pulse width modulation (PWM).) PCM involves sampling the analogue waveform at an appropriate rate, encoding the samples in digital (normally binary) form, and then transmitting the coded samples using a suitable digital waveform (which is also often binary but may well be ternary, quaternary or other). At the receiver the digital waveform is decoded, and the original message signal is reconstructed from the sample values. The complete process is illustrated in Fig. 7.1, where the transmitted digital signal has been shown as a baseband binary waveform for simplicity.

The most common application of PCM is perhaps in telephony, although the analogue message signal can originate from a wide range of sources other than a telephone handset: telemetry, radio, video, and so on. Similar techniques are also used for digital audio recording.

Later sections of the chapter introduce alternatives to standard PCM for the digital transmission of analogue signals, and also discuss the need for modems when using digital sources such as facsimile or computer terminals. Finally, some aspects of interfacing to an integrated services digital network ISDN) are introduced.

7.2 Sampling

The sampling rate used in a PCM system will depend on the bandwidth of the message signal. As discussed in Chapter 2, in order to avoid aliasing, the sampling rate must be greater than twice the highest significant frequency component of the message signal. A telephone speech signal is limited by filters to a range from about 300 Hz to 3400 Hz, and is usually

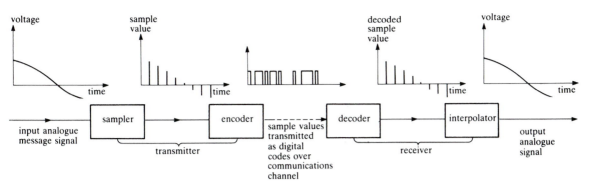

Fig. 7.1. The principle of pulse code modulation (PCM).

sampled at a frequency of 8 kHz for digital transmission. Using a sampling rate rather greater than the theoretical minimum is necessary for a number of practical reasons. For example, it enables the speech signal to be strictly bandlimited before sampling, using a practical anti-aliasing filter; similarly, it eases demands on the filters used to reconstruct the message signal at the receiver. CCITT recommendations include detailed specifications of the bandlimiting to be applied to a voice signal before sampling and encoding.

7.3 Encoding and decoding

Before coding, each sample of an analogue message signal must be allocated to one of a finite number of *quantisation levels*, in a way very similar to normal analogue to digital conversion. The process of sampling and quantisation together is often known as *digitisation*.

Quantisation is illustrated in Fig. 7.2, for a system using 4-bit binary words. There are thus $2^4 = 16$ different quantisation levels, each of which is allocated a different 4-bit code. For simplicity, the lowest level has been labelled 0000 and the highest 1111, although in practice rather different numbering schemes are used, as will be outlined below. Because of the finite number of levels, an analogue value cannot, in general, be encoded exactly, but must be rounded up or down to the nearest level as appropriate. If the levels are separated by q volts, then the maximum error introduced by the quantisation process is $\pm q/2$ – provided, of course, that the message signal itself does not exceed the top or bottom quantisation level by more than half an interval. The mean quantisation error is zero, as illustrated in part (b) of Fig. 7.2, which is obtained by subtracting the quantised signal from the message signal. Note that the figure shows only quantisation; sampling is ignored for the time being.

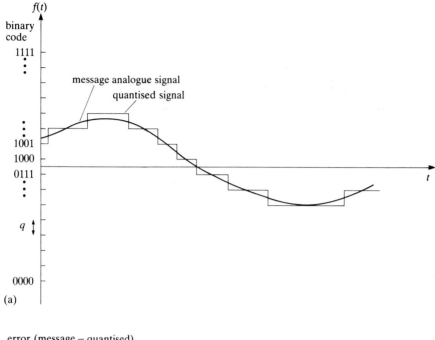

(a)

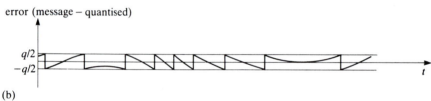

(b)

Fig. 7.2. The quantisation process using a 4-bit linear encoder.

Quantisation error sets a fundamental limitation to the performance of a PCM system, since even a signal perfectly reconstructed from the transmitted codes will suffer from it. The effect is usually known as quantisation noise or quantisation distortion, and is quantified in terms of the signal to quantisation noise ratio of the system.

7.3.1   Signal to quantisation noise ratio

The signal to quantisation noise ratio is usually defined as

$$S/N = \frac{\text{signal power}}{\text{quantisation noise power}}$$

As discussed in Chapter 3, it is usual to imagine both signal and noise powers being dissipated in a 1 ohm resistor, and hence use mean-square values as direct measures of the powers. Consider first the quantisation noise. To calculate the mean square we need to make some assumption about the probability distribution of the error. The simplest assumption is the distribution shown in Fig. 7.3: any value in the range $\pm q/2$ is equally likely. This model is found to hold well in practice providing the levels are sufficiently close. The height of the uniform distribution over this range is therefore $1/q$, in order to give a total area under the function of unity (the error must take on some value or other in the given range).

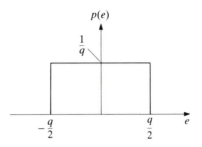

Fig. 7.3. Probability distribution of quantisation noise for a linear encoder.

The mean-square error or noise power $\overline{e^2}$ can now be calculated easily using the result of Chapter 3.

$$\overline{e^2} = \int\limits_{-\infty}^{+\infty} e^2 p(e)\, \mathrm{d}e$$

$$= \int\limits_{-q/2}^{+q/2} \frac{e^2}{q}\, \mathrm{d}e$$

$$= \left[\frac{e^3}{3q}\right]_{-q/2}^{+q/2}$$

$$= \frac{q^2}{12}$$

To calculate the signal power, some assumption must be made about the nature of the signal. The simplest assumption (although not a very realistic one) is that all values of the signal are equally likely over the permitted range of the encoder – that is, that the signal too has a uniform probability density function. The mean-square value of the signal can then be obtained exactly as above. The signal can take any value over a range of $2^n q$, where n is the number of bits used to encode a sample value (there are $2^n - 1$ quantisation *intervals*, plus the half interval above and below the top and bottom levels before the error exceeds $q/2$). Hence the limits of the integration become $\pm(2^n q)/2$ instead of $q/2$, and the mean square signal power can be deduced immediately to be $(2^n q)^2/12$. Thus:

$$S/N = \frac{\text{signal power}}{\text{noise power}} = \frac{(2^n q)^2/12}{q^2/12}$$

$$= 2^{2n}$$

Expressed in decibels, therefore

$$S/N = 10 \log 2^{2n} \, \text{dB}$$
$$= 20n \log 2 \, \text{dB}$$
$$\simeq 6n \, \text{dB}$$

EXAMPLE

Suppose that the signal is a sinusoid of the maximum value the encoder can accommodate without the error ever exceeding $q/2$. Obtain an expression in dB for the signal to quantisation noise ratio.

Solution. The amplitude of the sinusoid (half the peak-to-peak value) is $2^{n-1} q$ and its mean-square value is hence

$$\frac{(2^{n-1} q)^2}{2}$$

Assuming that the quantisation noise power is the same as before, the signal to noise power ratio becomes

$$\frac{S}{N} = \frac{(2^{n-1} q)^2/2}{q^2/12} = 1.5 \times 2^{2n}$$

In decibels, therefore

$$S/N = 10 \log 1.5 + 20n \log 2$$
$$\simeq (1.8 + 6n) \, \text{dB}$$

Note the slightly different expressions for S/N in the two cases modelled, because of the different assumptions made about the signal. The important point to note, however, is that the S/N improves by 6 dB for each additional bit in the wordlength of the encoder. This is because the number of intervals is doubled, the average error is halved, and the mean-square error is reduced by a factor of 4 – that is, by 6 dB.

The above expressions for signal to quantisation noise ratio have been derived by considering the quantisation process alone, as was shown in Fig. 7.2. Neither sampling (at the transmitter) nor signal reconstruction (at the receiver) has been modelled explicitly. After smooth signal reconstruction of the type discussed in Chapter 3 the quantisation error will no longer take the precise form of Fig. 7.2(b). Nevertheless, it can be shown that for sampling at the Nyquist rate followed by ideal reconstruction the previously derived results continue to hold. (If the signal is sampled faster than the Nyquist rate the quantisation noise can be reduced somewhat.) The full analysis of quantisation noise under these circumstances is quite complex, and will not be attempted here.

7.3.2   Non-linear encoding

So far it has been assumed that the quantisation levels are equally spaced. In practice, however, this leads to problems. The amplitude of a typical speech signal can vary enormously (30–40 dB) between quiet and loud passages. Furthermore, the variation from speaker to speaker is considerable – so much so, in fact, that PCM telephony has to cope with an overall dynamic range of as much as 50–60 dB. With equally-spaced levels, the signal to quantisation noise ratio will be worse for signals towards the bottom of the range than for those towards the top. To keep quantisation noise sufficiently low during the quietest tones of the quietest speakers, and still cope with the powerful tones of a much louder speaker, would require something like 2000 or more equally spaced levels – that is, 11-bit words or longer would be necessary.

Using such a large number of equally-spaced levels, though, is inefficient. The overall bandwidth required by the digital link will be proportional to the number of binary digits in each code word: the more digital symbols transmitted per sample, the faster the signalling rate, and hence the greater the bandwidth required. The solution adopted in practice is to use levels which are closer together for low-intensity signals and further apart for louder sounds. In this way the S/N can be kept reasonably constant whatever the level of the speech signal, using a modest number of bits per sample.

This technique is known as *non-uniform* or *non-linear encoding*, and is illustrated in Fig. 7.4 for a sinusoidal signal. Note how the spacing between levels becomes progressively wider as the signal level increases. In voice telephony, 8-bit non-uniform encoding is used to cover a range which would require 12 bits per sample with uniform quantisation. Two different encoder/decoder characteristics are specified by CCITT, known as *A-law* and *μ-law* respectively. Both are similar to the curve of Fig. 7.5, although the precise mathematical form differs in the two cases. In the figure, the V axis from -1 to $+1$ represents the full range of input voltages. The binary codes are imagined to be equally spaced along the

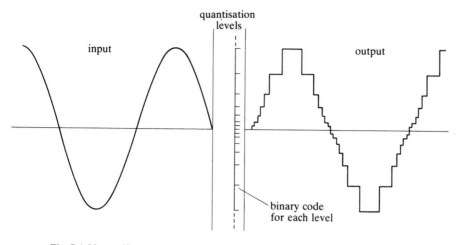

Fig. 7.4. Non-uniform encoding of a sinusoid. Smaller quantisation intervals are used
for small signal levels than for large levels.

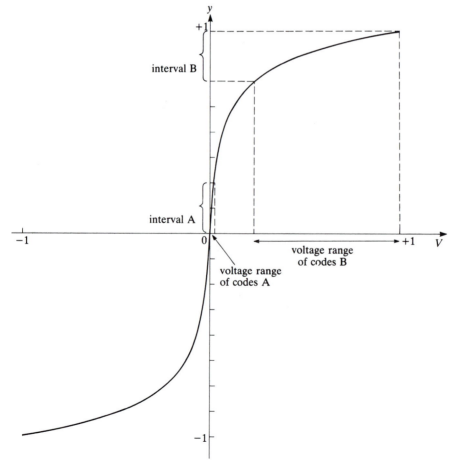

Fig. 7.5. Non-uniform encoder characteristic.

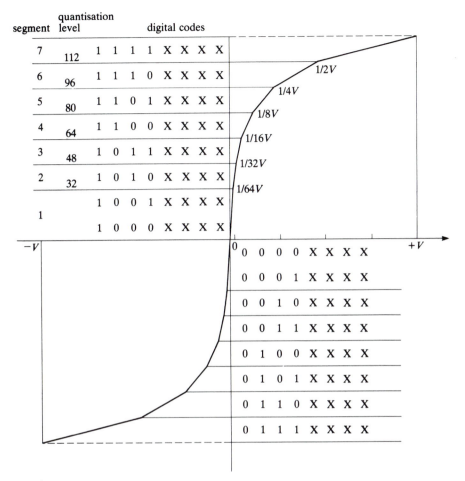

Fig. 7.6. Coding scheme for CCITT European piecewise linear approximation to A-law.

y-axis, so that the range of signal voltages represented by the codes in interval A near the origin is much smaller than that represented by an equal interval B near the maximum. The digital codes used for the 256 quantisation levels in the two schemes are listed in full in the CCITT recommendations, and commercial integrated circuits known as *codecs* (coder/decoders) are available for both A- and μ-law. European telephone systems use A-law, and North American and Japanese systems μ-law. Interconversation is performed where necessary for international telephone calls.

Fig. 7.6 gives details of the 8-bit, A-law piecewise linear characteristic used in European PCM. Note that there are 13 segments in all, since the first positive and negative segments form a single straight line. (These segments are also twice as wide as the others.) In the figure, the symbol

XXXX in an 8-bit codeword means that those particular bits vary from 0000 to 1111 within the segment. Note the numerical ordering of the binary codes. A leading 0 indicates a negative voltage, a 1 a positive voltage. The other seven bits 'count' away from zero in natural binary representation. The μ-law characteristic is very similar, except that a total of 15 linear segments are used, and the arrangement of binary codes is different: here the lowest natural binary numbers correspond to the largest negative quantisation levels.

As already mentioned, if the smallest quantisation interval in these schemes were used over the whole range, 12 bits per sample would be necessary. The effect of the non-linear encoding/decoding may therefore be thought of as 'compressing' a 12-bit to an 8-bit representation. For this reason the process is often known as companding (compressing/expanding). Both A- and μ-law companding characteristics are designed to give a signal to quantisation noise ratio which hardly varies over the whole of the range, as shown in Fig. 7.7 for an A-law device. For signals at the very bottom end of the range (segment 1) the encoder behaves very much like a linear encoder; the quantisation noise power is constant at $q^2/12$ so that the S/N falls off steadily with signal strength. Over the rest of the range, however, the logarithmic companding characteristic ensures a constant S/N. (Note that the decibel scale of the horizontal axis of Fig. 7.7 refers to a particular test signal specified by CCITT. The rapid deterioration of S/N at the very top end of the range occurs when the compander saturates – that is, the signal exceeds the maximum permitted level and is clipped.)

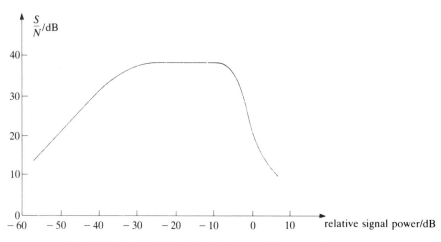

Fig. 7.7. Variation of S/N with signal strength for a non-linear encoder.

7.3.3   Errors in decoded PCM

The received sample value in a PCM system may differ from the original sample value not only because of quantisation noise, but also because of threshold detection error(s) caused by noise in the channel. The effect of such an error on the decoded sample value will depend on the bit(s) affected in the code word – clearly an error in the most significant bit will be more serious than one in the least significant, an effect which may be magnified by non-linear encoding. In general, quantisation noise is designed to be the limiting factor in a practical link. If the channel error rate becomes significantly high, the link is deemed to have failed, so the effect of such errors can be neglected except for quality control. The effect a single error may have is illustrated in the following example.

EXAMPLE

The transmitted code word 11111001 is received as 11110001 in a system using A-law encoding of signals in the range -5 to $+5$ V. What is the magnitude of the voltage error in the decoded word?

Solution. Referring to Fig. 7.6 it can be seen that the channel error has affected the topmost (positive) segment. The particular bit affected in this segment has a weighting $V/4$ for an encoder with a total range from $-V$ to $+V$. In this case, then, the magnitude of the voltage error introduced is equal to $1.25\,V$.

7.4   Signal reconstruction

Once the sample values have been decoded at a PCM receiver, the original analogue signal has to be reconstructed by interpolation between sample values, as described in Chapter 2. One way of doing this would be first to generate a train of short pulses with heights proportional to the individual sample values. From Chapter 2 the spectrum of such a pulse train (provided the pulses are sufficiently short) is a repeated version of the original message signal spectrum; as described earlier, ideal lowpass filtering would recover the message signal itself (apart from quantisation noise and any other errors introduced during transmission).

In practice, however, signal reconstruction is not normally carried out in quite this way. One difficulty is that the shorter the pulses (and they would need to be short, remember, to be treated as impulses, and hence give the perfect repeated spectrum), the weaker the signal at the output of

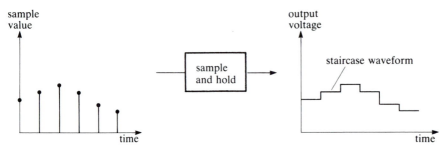

Fig. 7.8. Effect of 'holding' a sequence of sample values.

the lowpass filter. The most common approach in practice is to use a 'hold' device to produce a staircase waveform at the output of the decoder, rather than a pulse train. This is illustrated in Fig. 7.8; each sample can be thought of as being converted into a rectangular pulse width T and height proportional to the sample value. Such a process introduces new distortion, which needs to be counteracted by the reconstruction filter.

This distortion can be modelled most easily in the frequency domain. Let us compare the spectrum of an ideal short pulse or impulse (a sequence of which, when followed by ideal lowpass filtering at half the sampling frequency, would perfectly reproduce the message signal) with that of the 'full-width' pulse T seconds in duration. From Chapter 2, the amplitude spectrum of an impulse is flat, while that of a rectangular pulse takes the form

$$VT \left| \frac{\sin(\pi f T)}{\pi f T} \right|$$

Hence, if the full-width pulses, rather than impulses, are followed by an ideal lowpass filter with a cut-off of half the sampling frequency $1/2T$, the message signal will suffer the amplitude distortion illustrated in Fig. 7.9. To counter this, the ideal lowpass filter is replaced by one whose spectral

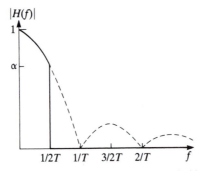

Fig. 7.9. Amplitude distortion introduced by the hold operation.

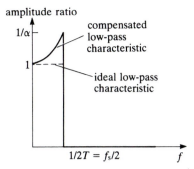

Fig. 7.10. Compensating for the amplitude distortion. Over the range $0 < f < 1/2T$ the amplitude ratio should approximate to $1/|H(f)|$, where $|H(f)|$ is as shown in Fig. 7.9.

characteristics approximate to Fig. 7.10, thus compensating for the '$(\sin x)/x$ distortion'.

EXAMPLE

Calculate the amplitude distortion, in dB, introduced over the range of a telephone speech signal by a recovery process consisting of a hold followed by ideal lowpass filtering without compensation. Assume a bandwidth of 300–3400 Hz and a sampling frequency of 8 kHz.

Solution. Given $|H(f)| = \left| \dfrac{\sin(\pi f T)}{\pi f T} \right|$, where $T = 125 \times 10^{-6}$ s

we need to calculate $\dfrac{|H(300)|}{|(3400)|} = A$, say.

Substituting in appropriate values gives $A = 1.37 \simeq 2.7$ dB

This 2.7 dB amplitude distortion introduced over the range of a telephone speech signal is a substantial contribution to the total permitted by CCITT standards for a PCM channel, and is normally corrected by suitable codec circuitry.

EXERCISE 7.1

A signal with components from d.c. to 1.5 kHz is sampled at a frequency of 4 kHz and reconstructed using a hold followed by an ideal lowpass filter with a cut-off of 2 kHz. System gain is such that a d.c. component is passed without attenuation. By how much (in dB) would a 1.5 kHz component be attenuated?

7.5   Systems aspects of PCM

This book is concerned primarily with the transmission of digital signals, rather with higher-level systems aspects of digital telecommunications. Nevertheless, any description of PCM would be incomplete without some reference to such systems aspects.

7.5.1   Codecs

Codecs were briefly mentioned above. A codec (encoder/decoder) can be implemented as a single integrated circuit which carries out anti-aliasing filtering, sampling, quantisation, and non-linear coding of the transmitted signal as well as decoding and signal recovery (including $\sin x/x$ equalisation) on the receive side.

The extract opposite is taken from a manufacturer's data sheet for a codec, and includes a simplified block diagram showing the main functional parts. The CCITT recommendations referred to are those mentioned above in Section 7.2, and specify the bandlimiting of voice signals prior to PCM encoding. Note that the bandlimiting is carried out by digital filters which have to operate at sampling rates much higher than 8 kHz to avoid the appearance of out-of-band input *analogue* components in the final *sampled* signal as a result of aliasing. Note also a particular feature of the μ-law codec, which improves timing content by ensuring that the code word 00000000 is never used.

7.5.2   Time-division multiplexing

As previously mentioned, an encoded telephone speech signal is transmitted at a rate of 64 k bits s^{-1} (8 bits per sample, 8000 samples per second). Long- distance telephone trunks can be designed to handle data at a much greater rate than this, however, and it is usual for a number of channels to share the same transmission link, using the technique of *time-division multiplexing*. Fig. 7.11 shows the basic principle, where the multiplexer and demultiplexer are represented by rotating switches. In telephony it is normal practice to use one encoder and decoder per channel, multiplexing on a sample-by-sample (octet-by-octet) basis, although bit-by-bit multiplexing is also possible. Multiplexing C channels, each sampled at S samples per second and coded using N bits per sample thus results in a composite bit rate of $C \times S \times N$ bits s^{-1}.

MV3506/7/7A/8/9

FUNCTIONAL DESCRIPTION

Fig.2 shows the simplified block diagram of the devices. They contain independent circuitry for processing transmit and receive signals. Switched capacitor filters provide the necessary bandwidth limiting of voice signals in both directions. Circuitry for coding and decoding operates on the principle of successive approximation, using charge redistribution in a binary weighted capacitor array to define segments and a resistor chain to define steps.

Transmit Section

Input analog signals first enter the chip at the uncommitted op.amp. terminals (IN + and IN- pins). This allows for the gain in the system to be trimmed. From the V_{IN} pin the signal enters a second-order analog anti-aliasing filter. This filter eliminates the need for any off-chip filtering as it provides attenuation of 34dB (typically) at 256kHz and 44dB (typically) at 512kHz.

The signal next enters the transmit filter, which is a fifth-order low-pass filter clocked at 256kHz, followed by a third-order high-pass filter clocked at 64kHz. The resulting bandpass characteristics meet the CCITT specifications G.711, G.712 and G.733. Some representative attenuations are better than 26dB from 0 to 60Hz and better than 35dB from 4.6kHz to 100kHz.

The output of the transmit filter is sampled at the analog to digital encoder by a capacitor array at the sampling rate of 8kHz. The successive approximation conversion process requires about 72μsec.

The 8-bit PCM data is clocked out by the transmit shift clock which can vary from 64kHz to 2.048MHz in 8kHz steps (see Figs. 3 and 4). A switched capacitor dual-speed, auto-zero loop using a small non-critical external capacitor (0.1μF) provides DC offset cancellation by integrating the sign bit of the PCM data and feeding it back to the non-inverting input of the comparator.

Included in the circuitry of the MV3507 is 'All Zero' code suppression so that negative input signal values between decision values numbers 127 and 128 are encoded as 00000010. This prevents loss of repeater synchronisation by DS1 (T1) line clock recovery circuitry as there are never more than 15 consecutive zeros.

An additional feature of the MV3506/7 and 7A is a special circuit to eliminate any transmitted idle channel noise during quiet periods. When the input of these chips is such that for 250ms the only code words generated were +0, -0, +1 or -1, the output word will be a +0. The steady +0 state prevents alternating sign bits or LSB from toggling and thus results in a quieter signal at the decoder. Upon detection of a different value, the output resumes normal operation resetting the 250ms timer. This feature is a form of idle Channel Noise 'Squelch' or 'Crosstalk Suppression'. It is of particular importance in the MV3506 A-Law version because the A-Law transfer characteristic has 'mid-riser' bias which enhances low level signals from crosstalk.

Receive Section

A receive shift clock, variable between the frequencies of 64kHz and 2.048MHz clocks the PCM data into the input buffer register once every sampling period (see Figs.5 and 6). A charge proportional to the received PCM data word appears on the decoder capacitor array of the digital to analog converter. A sample and hold circuit, initialised to zro by a narrow pulse at the beginning of each sampling period, integrates the charge and holds it for the rest of the sampling period.

The receive filter, consisting of a switched-capacitor fifth-order low-pass filter clocked at 256kHz, smooths the sampled and held signal. It also performs the loss equalisation to compensate for the sin(x)/x distortion due to the sampling.

The filter output (FLT OUT pin) is available for driving electronic hybrids directly as long as the impedance is greater than 20kΩ. When used in this fashion the low impedance output amp can be switched off for a considerable saving in power consumption. When it is required to drive a 600Ω load the output amp allows gain trimming as well as impedance matching.

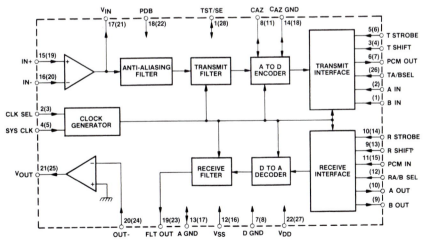

Fig.2 Functional block diagram (pin numbers for the MV3507A are in brackets)

Extract from manufacturer's data sheet for a codec.

Reprinted from Telecoms IC Handbook with permission of Plessey Semiconductors Ltd.

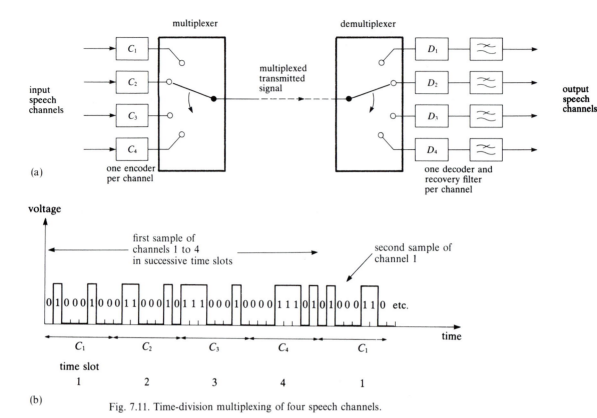

Fig. 7.11. Time-division multiplexing of four speech channels.

Fig. 7.11 shows four multiplexed channels, but in practice many more are multiplexed over a single link. The CCITT specifies a whole hierarchy of such multiplexing. At the lowest or primary level of the European system, 30 speech channels are multiplexed with the equivalent of two other channels used for the supervisory, timing and routing signals needed by the network: this is often known as the $30+2$ channel primary system. The bit rate of the primary multiplexed system is therefore $32 \times 8000 \times 8 = 2.048$ M bits/s. Higher levels will be discussed in the next chapter.

The time slots of the $30+2$ primary system are numbered from 0 to 31. Time slots 1–15 and 17–31 are used for the 30 speech channels, leaving slots 0 and 16 for the control information. The composite 256-bit digital signal of time slots 0–31 is referred to as a frame.

7.5.3   Timing and synchronisation

Maintaining the synchronisation of PCM transmission is a complex task, some aspects of which will be examined in more detail in the next chapter.

One or two important features are worth mentioning here, however. To begin with, recall that each receiver in the system has to be able to recover an accurate timing signal from the received bit stream – this procedure is known as clock recovery. To do this, there have to be sufficiently frequent transitions between signalling states, as discussed in Chapter 5 in connection with line codes. After a long enough period without such a transition it is possible for a receiver to get 'out of step' with the incoming bit stream. Preventing such long periods without a transition in signalling state is achieved by means of a combination of techniques. One simple one was noted above, in the description of a μ-law codec, where the 'all zero code' is simply not used, so that the worst possible case is 14 consecutive zeros, when the word 10000000 is followed by 00000001. A different technique known as alternate digit inversion, or ADI, is used in connection with A-law encoding (not to be confused with AMI). Here, every other digit in the code-word generated by the codec is inverted. Hence, if a steady string of zeros is generated, as will be the case in the idle state, the sequence actually transmitted is ... 10101010101 ... (Note, of course, that if the codec generates the sequence ... 10101010 ... then this will become all zeros after ADI. This particular sequence, however, is sufficiently unlikely in coded speech to be neglected.) These rudimentary techniques are adopted in addition to more sophisticated line coding as discussed earlier in Chapter 5.

Loss of synchronisation would be caused by a 'bit slip' anywhere in transmission. This is the reason for taking special measures to avoid bit slips, as mentioned in various places throughout the book.

Suitable line coding can ensure synchronisation at the bit level over long periods. Suppose that a timing error does occur, however, for some reason. Without some way of re-synchronising the system so that individual PCM frames and ultimately octets can be identified, the digital data cannot be correctly decoded. The use of the equivalent of two channels in each $30+2$ channel primary group for signalling and synchronisation information has already been mentioned. Time slot 0 is used for frame alignment, by arranging for alternate frames to carry octets X0011011 and X10XXXXX in this particular slot (X indicates that the bit may be either a 1 or a 0). Once the beginning of a frame is identified by recognising such a *frame alignment word (FAW)*, the receiving system can count the bit positions to identify the time slots, assuming the extraction of an accurate clock signal. Because any frame alignment word will occasionally arise in coded speech or computer data, two alternate words are used and the presence–absence–presence cycle of the two FAWs is monitored to prevent abortive attempts to synchronise with a chance bit pattern.

The situation is, in fact, even more complex than this. PCM frames are often grouped into larger, multi-frames each consisting of 16 individual frames. In frame 0 of the multi-frame, time slot 16 holds a *multi-frame*

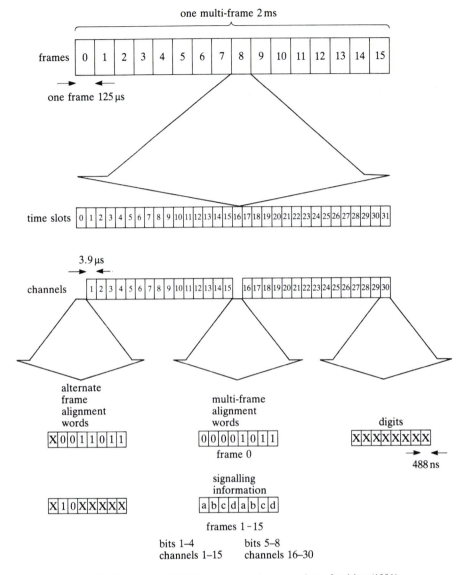

Fig. 7.12. Thirty-channel PCM structure common at time of writing (1991).

alignment word 00001011. (In the other 15 frames time slot 16 carries the signalling information used for setting up and clearing calls, details of which are beyond the scope of this text.) The situation is summarised in Fig. 7.12; further information can be found in CCITT recommendation G.704.

Although still common at the time of writing, the use of multi-frames will become unnecessary as a different technique for sending signalling information, known as *common channel signalling*, is introduced.

7.6 Delta modulation and differential PCM

Many message signals encountered in practice change only slightly from one sampling instant to the next – at least for the vast majority of the time. In such cases it is extravagant to encode the *absolute* value of each sample: transmitting information about the *change* from the previous sample instead would require considerably fewer bits. The simplest way of doing this is to transmit one bit of information per sample, in order to specify whether the signal is increasing or decreasing. This technique is known as *delta modulation*, and the essentials of a transmitter and receiver are shown conceptually in Fig. 7.13. Under normal operating conditions the feedback loop in the modulator (part (a) of the figure) forces the integrator output $\hat{m}(t)$ to follow the message signal closely: $\hat{m}(t)$ is thus an approximation to, or an *estimate* of the message signal. The sampled output $d(t)$ of the comparator consists of a (short) positive pulse if the message signal is greater than the estimate, and a negative pulse if it is smaller; $d(t)$ can therefore be thought of as transmitting information about the change in the signal from one sampling instant to the next. Waveforms are illustrated in Fig. 7.13(b). Note how the pulses $d(t)$, when integrated in the feedback path, form the stepped approximation $\hat{m}(t)$ of the message signal $m(t)$.

At the receiver, also shown in Fig. 7.13(a), $d(t)$ is again integrated to give the stepped approximation $\hat{m}(t)$ to the message signal. While $m(t)$ is increasing with time, $d(t)$ consists of more positive pulses than negative, and the output of the receiver integrator also tends to increase; conversely, as $m(t)$ decreases, so will the integrator output. While the message signal $m(t)$ is unchanging, the delta modulator is designed to generate an alternating idling sequence of positive and negative pulses. The final stage in the receiver is simply to smooth (lowpass filter) the stepped waveform $\hat{m}(t)$ so that the receiver output closely resembles the original message.

In a practical delta modulator $d(t)$ need not take the form of the short pulses shown. Full-width pulses might be used, for example, and the integration approximated by a simple RC-circuit. Neither would $d(t)$ be transmitted directly. Typically, the positive and negative pulses would be represented by a binary sequence of zeros and ones, and the bit stream transmitted over a digital channel in the usual way.

Two particular features of delta modulation, both shown in Fig. 7.13(b), should be noted. The first is that there is a start-up period during which time the feedback loop 'locks in' to give close agreement between message signal $m(t)$ and estimate $\hat{m}(t)$. During normal operation, the

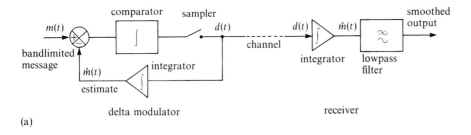

(a)

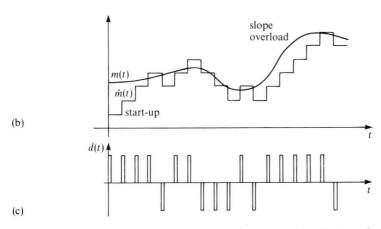

(b)

(c)

Fig. 7.13. Delta modulation showing transmitter structure, idealised waveforms and receiver structure.

output of the receiver closely follows the message signal, but if the latter changes too quickly, the delta modulator cannot keep up and *slope overload* occurs. The maximum slope which can be followed accurately will depend on the sampling rate chosen and the step size. In order to keep quantisation distortion low and prevent slope overload, delta modulation requires a sampling rate much greater than that of PCM for the same signal. For certain applications, though, this is more than compensated for by the simplicity of the circuitry involved, and the need to transmit only one bit rather than, say, eight per sample. A further advantage of delta modulation is that a channel error has only a small effect on the decoded output. In standard PCM a channel error in one of the more significant bits can give rise to substantial output error, as noted earlier.

EXAMPLE

(a) A delta modulator uses a step size σ and a sampling frequency f_s. Derive an expression for the maximum amplitude sinusoid of angular frequency ω which can be transmitted without slope overload.

(b) What is the smallest message amplitude (threshold level) which will disturb the idling pattern of alternating pulses?

Solution. (a) The maximum slope which can be followed is σf_s, (corresponding to a 'staircase' integrator output). The maximum rate of change of the message sinusoid $m(t) = A \cos \omega t$ is at the zero crossings, where, by differentiating, it can be seen to equal $\pm \omega A$. Hence slope overload occurs if $A > \sigma f_s / \omega$.

(b) Variations in $m(t)$ smaller than σ cannot be detected. Hence a sinusoid of amplitude greater than $\sigma/2$ (σ peak-to-peak) is needed to disturb the idling pattern.

A problem with delta modulation as just described is that a small step size σ is needed to reduce quantisation noise and threshold effects (granularity), but a large step size to limit slope overload. There are a number of ways of overcoming this apparently conflicting requirement. In *differential* PCM (DPCM) the change in the message signal is represented by a code word using *several* bits per sample rather than the *one* of delta modulation. Considerably fewer bits per sample therefore need to be transmitted than in standard PCM, yet the 'coarseness' of simple delta modulation is avoided. The drawback, however, is that the hardware is considerably more complicated – for example, analogue to digital conversion is now required, in contrast to the single comparator of the simpler delta modulator. An alternative approach, known as *adaptive delta modulation*, is similar to the companding technique introduced earlier. In adaptive delta modulation the step size is varied according to the rate of change of the message signal. If the slope is large, the step size is increased automatically to prevent overload; if the slope is small the step size is reduced to decrease the threshold level and the quantisation noise. In some systems a range of discrete step sizes is available; in others, the step size can be varied continuously (*continuously variable slope delta modulation* or *CVSDM*). Adaptive DPCM is also sometimes employed.

A simple variant of delta modulation is known as *delta–sigma modulation*. Suppose that the integrator is removed from the receiver (demodulator) of the standard delta modulation system of Fig. 7.13 and applied to the message signal instead, as shown in Fig. 7.14(a). Block diagram manipulation then leads to the equivalent form of modulator shown in Fig. 7.14(b). Such delta–sigma modulation has a number of advantages. For example: d.c. components can now be transmitted; the receiver consists simply of a lowpass filter; and channel errors are no longer cumulative (since the receiver does not integrate them).

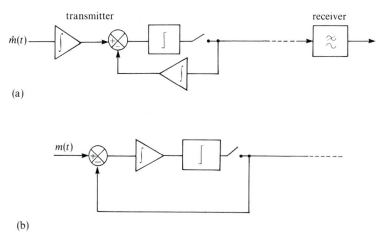

Fig. 7.14. Delta-sigma modulation: (a) 'moving' the receiver integrator to the transmitter and (b) new transmitter structure.

7.7 Digital sources

Although the bulk of telecommunications traffic is speech, there is an increasing demand for the communication of digital data: for example, computer communications or facsimile (fax) signals. Packet switched networks exist specifically for carrying digital information at high bit rates, but since they are not as widely accessible as the PSTN, there remains a large demand for sending digital data over the PSTN. Since switching and trunk transmission in the PSTN is almost entirely digital, it would seem a straightforward matter to transmit digital data. This is, indeed, the principle behind ISDN (discussed below). Unfortunately, as outlined in the introduction to Part 3 above, most links from local exchanges to private customers still use analogue transmission and, because of the numbers involved, many such links will remain analogue for some time to come. Consequently, the conversion of signals from digital sources into a form suitable for transmission over analogue speech circuits is likely to be required for the foreseeable future. The device which performs this conversion is known as a modem (*mod*ulator/*dem*odulator).

7.7.1 Modems

Table 7.1 lists some of the CCITT specifications for modems for use on the PSTN, taken from the Blue Book 'V series' recommendations. All those shown here are for use by dialling through the public switched network.

Packet switched networks, such as British Telecom's PSS network, are beyond the scope of this book. A useful introduction will be found in Block 2 of the Open University course T322.

The overheads of the signal conversions in a modem and a codec lead to an enormous inefficiency when transmitting inherently digital data in this way over the PSTN. A common occurrence will be for a modem to convert a 300 bit s⁻¹ signal into a stream of sinusoidal pulses, using one of the techniques described in Chapter 4. These sinusoidal pulses are then sampled and converted to a 64 kbit s⁻¹ digital signal by a codec at the local exchange!

Table 7.1. *Some of the modems specified by CCITT for use over PSTN switched circuits*

CCITT specification number	Forward channel	Backward channel
V.21	300 bits s^{-1} FSK signalling frequencies 980 and 1180 Hz	300 bits s^{-1} FSK signalling frequencies 1650 and 1850 Hz
V.22	1200 bits s^{-1} 4-DPSK carrier 1200 Hz (scrambler $1+x^{-14}+x^{-17}$)	1200 bits s^{-1} 4-DPSK carrier 2400 Hz (scrambler $1+x^{-14}+x^{-17}$)
V.22 bis	2400 bits s^{-1} QAM carrier 1200 Hz (scrambler $1+x^{-14}+x^{-17}$)	2400 bits s^{-1} QAM carrier 2400 Hz (scrambler $1+x^{-14}+x^{-17}$)
V.23	1200 bits s^{-1} FSK signalling frequencies 1300 and 2100 Hz	75 bits s^{-1} FSK signalling frequencies 390 and 450 Hz
V.26	2400 bits s^{-1} 4-DPSK carrier 1800 Hz (scrambler $1+x^{-18}+x^{-23}$)	as forward, but with scrambler $1+x^{-5}+x^{-23}$
V.32	9600 bits s^{-1} QAM carrier 1800 Hz (scrambler $1+x^{-18}+x^{-23}$)	as forward, but with scrambler $1+x^{-5}+x^{-23}$

CCITT also include specifications for use over *leased lines*. Leased lines are channels which are permanently linked (for the duration of the lease) by the same physical connection between two points. A modem used with a leased line can be matched to the characteristics of that particular line by setting up equalisation parameters on installation. By contrast, a modem used with a 'dial-up' line will be faced with different characteristics for each call. In general, telephone channels introduce a substantial degree of both amplitude and phase distortion, so modems will always need to apply some equalisation to the received signal before recovering the digital message. For the lower rate modems a simple fixed 'compromise' equaliser would be sufficient, but for greater rates, more sophistication is required. Commonly equalisation will be performed digitally, using techniques such as that described in Chapter 4.

FSK is used at the lower data rates because of its simplicity and because, unlike ASK, it provides a constant signal level regardless of the data sequence. The more bandwidth-efficient schemes of PSK or QAM

Table 7.2. *A DPSK modulation scheme used by V.22*

Input dibit*	Phase change
00	$+90°$
01	$+0°$
11	$+270°$
10	$+180°$

*Dibit just means two bits

are required at higher data rates. For the very highest data rates, error correcting channel coding is used in conjunction with the modulation. For example, one of the modes of operation for V.32 is with *trellis code modulation (TCM)*. This is a combination of convolutional coding and QAM. Because of the redundancy of convolutional coding, more states are required in the TCM QAM constellation than with QAM alone. Although this means that the states are closer together, with a resulting increase in probability of confusion, the benefits of being able to correct errors more than offsets the increase in error probability. The result is an overall decrease in error probability in the decoded data, for a given signal to noise ratio.

This reduction in error probability by the use of error correction coding is known as *coding gain*.

The PSK and QAM schemes are used differentially, rather than directly, because of the problem of phase synchronisation discussed in Chapters 4 and 5, and also to help maintain timing information. The latter aspect can be appreciated by considering the modulation scheme used by V.22 given in Table 7.2.

The receiver will derive symbol interval timing from the phase changes, so it is desirable to prevent long runs occurring with no change of phase. If four absolute phase states were used for the modulation states representing four dibits, then if any of those dibits were continuously transmitted (including the common 00...) there would be no change of phase in the output. In the scheme of Table 7.2, however, the only circumstance in which there will be no change in the output phase is if the data consists of the continuously repeating dibit 01...

V.26 goes further: the modulation scheme (Table 7.3) provides a phase change for any input dibit.

Duplex operation The lower rate systems (V.21, V.22 and V.22bis) provide the full data rate for both forward and backward channels by

Table 7.3. *A modulation scheme for V.26*

Input dibit	Phase change
00	$+45°$
01	$+135°$
11	$+225°$
10	$+315°$

using different signalling frequencies in the two directions. In other words they provide full duplex operation by frequency-division multiplexing.

At higher transmission rates, the full bandwidth is required for each direction of transmission. In this case full duplex operation can be achieved by either using a separate channel for each direction (four-wire operation) or by hybrid separation – the same technique as is used to separate the incoming and outgoing signals (the earpiece and mouthpiece signals) in an analogue telephone. The separation achievable with a hybrid alone, however, is inadequate for high-speed digital modems, so a feedback technique known as *echo cancellation* is also used. Basically, echo cancellation operates by identifying any components in the incoming signal which are the same as the outgoing signal and, assuming them to be due to unwanted leakage across the hybrid, subtracting them from the incoming signal.

7.7.2 ISDN

In many ways the ideal communication network is one in which the signal from each customer is digital, and remains digital through to its destination. Analogue sources (such as telephony) are coded at the customer's premises by PCM – or, indeed, any other suitable digital coding, such as delta modulation, specified in the relevant standards. This is the concept of the *Integrated Services Digital Network* (*ISDN*). CCITT have recommended standardised interfaces for customer access to ISDN (the *User Network Interface* or *UNI*).

The *basic rate* for ISDN access is two $64 \, \text{kbits s}^{-1}$ bearer channels (B channels) plus one $16 \, \text{kbits s}^{-1}$ data channel (D channel): the $2B+D$ access. The B channels carry services while the D channel carries the signalling information necessary for the control of services. In telephony, for example, a B channel would carry the telephone signal while the D channel would transmit dialling and other information necessary to set up and clear calls. However, the D channel can also carry some user packet data in those intervals when signalling data does not need to be transmitted.

The combined data rate, $144\,\text{kbits s}^{-1}$, plus overheads, is about the limit of what can be conveyed over most links between private customers and the local exchange. (These are the links which were installed to carry analogue telephone signals, and are generally pairs of twisted copper wires.) The CCITT (G961, Appendices I–VI) has suggested coding schemes which might be used for this link: 3B2T, 4B3T, or 2B1Q for example. (2B1Q has already been selected for this purpose in the USA.) Because the technology was not initially available to provide the full basic rate access however, some telecommunications authorities began implementing ISDN with lower rates. For example, British Telecom has used a rate of $80\,\text{kbits s}^{-1}$, which consists of $1B+D$ (plus overheads).

It should be noted that CCITT recommendations also specify *primary access* to the ISDN at data rates corresponding to the primary level of the US and European PSTN PCM hierarchies.

B-ISDN With the continuing information technology revolution, increasing numbers of services are appearing which require higher data rates than is provided by basic rate ISDN. The principal broadband services are those involving images and high-speed data transmission: digital video for video-telephony, image retrieval from video libraries, etc. This has led to the introduction of standards for the *Broadband Integrated Services Digital Network*: *B-ISDN*.

Some aspects of B-ISDN are radically different from ISDN. Specifically, end-to-end communication is by *asynchronous transfer mode*, (*ATM*), rather than by synchronous transfer mode (STM) as in ISDN. In STM, an identifiable channel exists over the entire link, dedicated to that link. Even where the link uses a multiplexed signal, defined bits in the multiplex frame will be given over to that link. The synchronous aspect can be thought of as the synchronisation between the incoming data rate and the transmission rate, in order to 'slot' the data bits into the transmitted signal.

By contrast, ATM operates by putting the data into short, fixed-length packets called *cells*, which are conveyed to the destination by the network. Each stage of a link can operate asynchronously: so long as all the cells reach their destination in the correct order, it does not matter at what precise bit rate they were carried.

The cell structure is shown in Fig. 7.15. It has a total of 53 bytes, consisting of 5 header bytes and 48 information bytes. One of the main functions of the header is to hold the *virtual channel* and *virtual path* identifiers (*VCI* and *VPIs*). More will be said about these in the following chapter.

ATM can be viewed as a type of packet switching. Previous packet

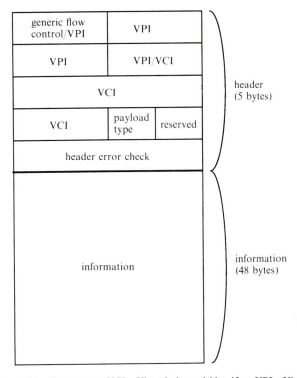

Fig. 7.15. ATM cell structure. VCI = Virtual channel identifier. VPI = Virtual path identifier. The header error check covers the header only. Cell overheads do not include an error check on the information. The header structure differs slightly at different points in the network. Specifically, the generic flow control is present only at the interface between the user and the network (the User-Network Interface: UNI), not at interfaces within the network (Network Node Interfaces: NNI).

switching networks were not suitable for conveying 'real-time' services such as telephony or live video, because packets can be delayed in transit while waiting for access to a link, and this would corrupt the received signal. However, ATM overcomes this limitation by identifying different types of connection, and guaranteeing for those connections requiring a continuous bit stream sufficient capacity to ensure that an excessive delay is not encountered.

7.8 Summary

In pulse code modulation an analogue waveform is sampled, quantised and encoded digitally for transmission. Sampling must be carried out at a rate greater than twice the analogue signal bandwidth to avoid aliasing. The quantisation process introduces errors known as quantisation noise or distortion. Assuming that any value of error in the range $\pm q/2$ is equally likely, where q is the quantisation interval, linear quantisation introduces mean-square quantisation noise of $q_2/12$. Providing that q is small compared with the signal, and the signal extends over the whole range of the quantiser, the signal to quantisation noise ratio is of the order of $6n$ dB, where n is the quantiser wordlength.

Current PCM telephony involves sampling a 300–3400 Hz analogue signal at a rate of 8000 samples s^{-1} followed by 8-bit, non-linear encoding, resulting in a 64 k bit s^{-1} digital signal. Non-linear quantisation (using an A-law characteristic in Europe and a μ-law characteristic in North America and Japan) ensures an approximately constant signal to quantisation noise ratio over a wide range of signal levels.

PCM encoding and decoding is performed by a codec, an integrated circuit which carries out the sampling and quantisation processes for the transmitted signal, and the reconstruction process for the received signal. The latter normally involves first the generation of a 'staircase' waveform as the output of a sample-and-hold device, followed by equalisation to compensate for the $(\sin x)/x$ distortion introduced.

Because the bandwidth of modern transmission channels is much wider than needed to transmit a single PCM telephone signal, such signals are time-division multiplexed in a hierarchy. At the lowest (primary) level 30 speech channels are multiplexed with the equivalent of two other channels used for supervisory, timing and routing signals: this is often known as the $30 + 2$ channel primary system and has a bit rate of 2.048 M bits s^{-1}.

A technique closely related to PCM is that of delta modulation, in which only one bit per sample is transmitted, according to whether the encoded message signal is increasing or decreasing. Delta modulation schemes are often made to adapt to the rate at which the message signal is changing in order to avoid the problems of slope overload and excessive granularity.

Although public switched telephone networks are increasingly using digital trunks, most links from local exchanges to private customers are still analogue. Modems are therefore necessary to convert signals from digital sources (computer data and fax, for example) into a form suitable for transmission over analogue speech circuits; CCITT recommendations cover modems operating at various rates using FSK, QAM and differential PSK. In an Integrated Services Digital Network, on the other hand, the signal from each customer is digital, and remains digital through to its destination. The proposed basic ISDN access consists of two bearer channels at 64 kbits s^{-1} together with one data channel at 16 kbits k^{-1} (2B + D), which fits in well with existing digital traffic over the PSTN. Broadband ISDN, which will provide for the much higher data rates needed for video, will operate in a very different way known as Asynchronous Transfer Mode (ATM). This can be thought of as a type of very fast packet switching in which real-time services can be guaranteed delays not exceeding a given level.

8

Digital transmission hierarchies

8.1 Introduction

In chapter 7 reference was made to the multiplexing of $2048\,\mathrm{kbits\,s^{-1}}$ channels to exploit the high capacity of modern transmission media. Over any given route, the most cost-effective transmission will, in general, be achieved by multiplexing as many as possible of the channels together. As technology has advanced, it has become feasible to combine an increasing number of channels by operating at ever higher signalling rates. Some routes, of course, will only need a few channels, so will not require very high signalling rates. Transmission networks in the public switched telephone network, therefore, are designed around hierarchies of transmission rates, corresponding to increasing numbers of channels conveyed on a single multiplexed link. These hierarchies are defined in national and international standards. The first hierarchies to be standardised were those for plesiochronous multiplexing (see Section 8.2 below). Unfortunately several distinct plesiochronous hierarchies simultaneously exist in different locations around the world. In Europe the hierarchy is based upon the 30-channel $2048\,\mathrm{kbit\,s^{-1}}$ primary rate, while in the USA and Japan it is based upon the 24-channel $1544\,\mathrm{kbits\,s^{-1}}$ primary rate. Furthermore, although the US and Japanese hierarchies are the same at lower levels, they diverge at higher levels.

For reasons which are discussed below, more recently it has been decided to move to synchronously multiplexed transmission hierarchies. There has been a concerted attempt to ensure that a multiplicity of new hierarchies does not arise. The first standards for a synchronous hierarchy were the Synchronous Optical Network (SONET) standards of the American National Standards Institute (ANSI) in the USA. These were taken up (but modified) by CCITT to become the world standard known as Synchronous Digital Hierarchy (SDH). Although there are differences between SONET and SDH, they remain largely compatible. Furthermore, they are designed to interface with both the $2048\,\mathrm{kbits\,s^{-1}}$ and $1544\,\mathrm{kbits\,s^{-1}}$ plesiochronous hierarchies; they are designed to 'draw together' the separate hierarchies.

In other countries of the world, the hierarchy used often depends upon where the majority of the telecommunications equipment was bought. Europe, the USA and Japan all have their own 'spheres of influence'.

Table 8.1

Digital hierarchy level	Rate, in kbits s^{-1}	
	US hierarchy	European hierarchy
1	1 544	2 048
2	6 312	8 448
3	44 736	34 368
4		139 264

8.2   The plesiochronous digital hierarchy (PDH)

Some of the successive levels of multiplexing recommended by CCITT are listed in Table 8.1. The structure of the European hierarchy is shown in Fig. 8.1, and the way it is used is illustrated in Fig. 8.2. Notice that the 34 Mbits s^{-1} transmission system (using a single pair of cables) is carrying four 8 Mbits s^{-1} channels. These carry a total of sixteen 2 Mbits s^{-1} channels, which in turn carry four hundred and eighty 64 kbits s^{-1} channels. The interconnections between blocks in Fig. 8.2 (between successive multiplexers or to and from transmission systems) are standardised interfaces, defined by CCITT in recommendation G.703. Examples of some parameters for these interface standards are summarised in Table 8.2. The signal structure is also standardised at these interfaces, the structure being determined by the details of the multiplexing procedures (discussed below). Standardisation of the interfaces allows telecommunications authorities to build and re-configure the transmission network on a modular basis, with different modules purchased from different suppliers.

Table 8.2 includes the bit-rate tolerances for each level. To maintain the output signal within these tolerances, each multiplexer contains a

As an international body, CCITT describes both the 2048 kbits s^{-1} based hierarchy and the 1544 kbits s^{-1} hierarchy. In principle, CCITT recommendations are intended to cover telecommunications practice throughout the world. Regional authorities however also have their own local standards which vary in the extent to which they correspond to the CCITT recommendations. In the USA, the regional standards are defined by ANSI.

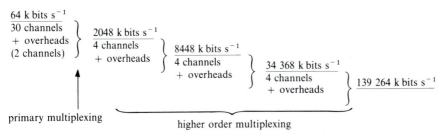

Fig. 8.1. European plesiochronous multiplexing structure.

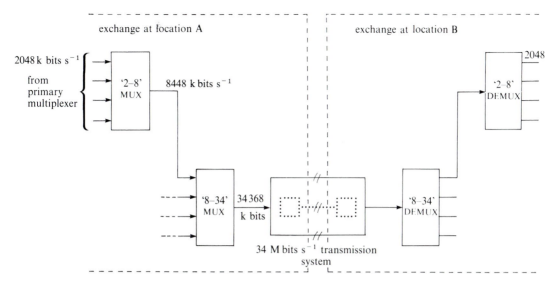

Fig. 8.2. Example of the plesiochronous digital hierarchy using transmission at 34 Mbits s⁻¹. (MUX is an abbreviation for multiplexer and DEMUX for demultiplexer).

In some countries the digital transmission network has been synchronised by distributing a clock signal around the country, to be used as the timing reference for all multiplexers. Although this does have some advantages, the full benefits of a synchronised network are only available if the *multiplexing* is made synchronous as well, as described in Section 8.3 below.

The term plesiochronous is not used in the USA. Plesiochronous signals are simply described as asynchronous. Also in the USA, justification is often referred to as stuffing: this should not be confused with the (very different) process of stuffing in X.25 link-level standards.

stabilised clock source. At a primary multiplexer, a subdivision ($1/32$) of the multiplexer's clock is also sent out to the codecs, so the timing of the 30 incoming $64\,\mathrm{k\,bits\,s^{-1}}$ channels is synchronised to the outgoing $2048\,\mathrm{Mbits\,s^{-1}}$ signal.

At higher order multiplexers, however, the four incoming ('tributary') signals will have originated at lower order multiplexers, each with its own clock source. So the four tributaries will each have a different precise bit rate – although all should be within the specified tolerance. Such signals, having nominally the same frequency but differing within a defined tolerance, are described as *plesiochronous* signals. Multiplexing plesiochronous signals obviously involves a synchronisation problem; this is overcome by the use of *justification* or *stuffing* as discussed below.

8.2.1   Multiplexing details

Primary multiplexing to form the $2048\,\mathrm{kbits\,s^{-1}}$ signal (as described in Chapter 7) operates by interleaving 8-bit bytes from each of the 30 channels, putting one byte from each channel in a frame. The frame structure is identifiable within the bit stream by the presence of the frame alignment word (FAW). Higher order multiplexing also uses a frame structure, bounded by a frame alignment word, but the multiplexing is bit-by-bit. Consequently the byte structure of the primary multiplexed

Table 8.2

Level	Bit rate[1] (kbits s^{-1})	Line code[2]	Peak pulse voltage	Maximum cable loss[3]
1	2 048 ± 50 ppm	HDB3	2.37 or 3.00 V[4]	6 dB
2	8 448 ± 30 ppm	HDB3	2.37 V	6 dB
3	34 368 ± 20 ppm	HDB3	1.00 V	12 dB
4	139 264 ± 15 ppm	CMI	1.00 V[5]	12 dB

Notes: (1) ppm stands for 'parts per million', so, for example, the tolerance on the level 1 signal is ± 102.4 Hz.
(2) These line codes are discussed in Chapter 5.
(3) The cable attenuation is assumed to follow an approximate $\sqrt{f}$ law, and the attenuation specified is for a frequency numerically equal to half the signalling rate (e.g. 1024 kHz for level 1).
(4) Level 1 can use either a coaxial cable or symmetrical pair. The pulse voltage should be 2.37 volts on coaxial cable, 3.00 V on symmetrical pair.
(5) This is the peak-to-peak voltage.

signal is not preserved at higher levels; the higher levels merely treat the signal as an unstructured 2048 kbits s^{-1} bit stream.

The methods used in higher-order plesiochronous multiplexing are conveniently described by looking at a specific example. Fig. 8.3 shows the multiplex frame structure for an 8448 kbits s^{-1} signal, as described in G.742. There are 848 bits in a frame, as follows:

10 bits	the frame alignment word (1111010000).
2 bits	a signalling overhead (one is used to indicate an alarm condition, the other specified as 'for national use').
820 bits	data from the tributaries. (205 from each tributary, in four blocks of sizes: 50, 52, 52, 51).
16 bits	used in the justification process (12 justification control bits, 4 justification opportunity bits).

The justification bits are used to match the precise signalling rates of the tributaries to the signalling rate of the multiplexed signal. Consider a single tributary (say number 1). With 205 bits from this tributary being carried in the multiplexed frame, the precise data rate being carried will depend upon the precise signalling rate of the multiplexed signal. From Table 8.2, the tolerance on the 8448 kbits s^{-1} signal is ± 30 ppm, so the range is 8 447 747 bits s^{-1} to 8 448 253 bits s^{-1}. The maximum data rate

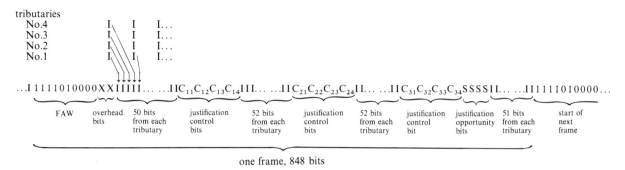

tributaries
No.4 I I I...
No.3 I I I...
No.2 I I I...
No.1 I I I...

...I 1 1 1 1 0 1 0 0 0 0 X X I I I I I.... ...IIC₁₁C₁₂C₁₃C₁₄III... ...IIC₂₁C₂₂C₂₃C₂₄II... ...IIC₃₁C₃₂C₃₃C₃₄SSSSII... ...II1111010000...

| FAW | overhead bits | 50 bits from each tributary | justification control bits | 52 bits from each tributary | justification control bits | 52 bits from each tributary | justification control bit | justification opportunity bits | 51 bits from each tributary | start of next frame |

one frame, 848 bits

I = information bit
C = justification control bit
S = justification opportunity bit
X = overhead bit
0,1 = bits with logic levels 0 or 1 respectively

Fig. 8.3. Possible multiplexed frame structure at $8448\,\mathrm{kbits\,s^{-1}}$.

from tributary 1 will therefore be $(205/848) \times 8\,448\,253$ bits s^{-1} = $2\,042\,325$ bits s^{-1}. The data rate arriving from the tributary, will be (from Table 8.2 again) 2048 kbits s$^{-1} \pm 50$ ppm, which is the range $2\,047\,898$ bits s^{-1} to ± 5 $2\,048\,102$ bits s^{-1}. Thus, whatever the precise signalling rates of either the tributary or the multiplexed signal (provided they are both within the specified tolerances), the data rate carried by the 205 'tributary number 1 data bits' in the multiplexed signal will be less than the data arriving from tributary number 1.

There will be a buffer in the multiplexer which will temporarily store the incoming tributary data, but clearly if the data is coming in faster than it is going out, the buffer will sooner or later overflow. However, the multiplexed frame contains an extra bit which can be used for carrying data from tributary number one; this is the first of the justification opportunity bits, bit S_1. Thus, by allowing bit S_1 to be used to carry tributary number 1 data, the outgoing data rate is increased from 205 to 206 bits per frame.

EXERCISE 8.1

With 206 bits from tributary one in the multiplexed signal, calculate the range of rates conveyed from tributary one in the multiplexed signal, for the tolerance on the 8448 kbits s^{-1} signal given in Table 2. Compare this with the range of incoming signalling rates for tributary one.

The calculation in Exercise 8.1 shows that at 206 bits per frame, the outgoing data rate for tributary number 1 will always be greater than the incoming data rate – again provided that the tributary data rate and the multiplexed data rate are both within their specified tolerances. Thus the output data rate can be made greater or lower than the input data rate depending upon whether S_1 is used to carry data or not (if S_1 is not carrying data, its value is irrelevant). By using S_1 in some frames and not in others, the output data rate averaged over many frames can be matched exactly to the incoming tributary data rate. Obviously, the demultiplexer needs to know whether or not the S_1 bit is carrying data. This information is encoded in the justification control bits. For tributary number one, the relevant bits are C_{11}, C_{21} and C_{31}. If S_1 is not being used, C_{11}, C_{21} and C_{31} are all 0, if S_1 is being used, C_{11}, C_{21} and C_{31} are all 1. Three bits are used rather than just a single bit in order to provide protection against errors. At the receiver a majority decision is taken on the three bits.

Justification is carried out independently on each of the four tributary channels through the use of the four sets of justification control bits and the four justification opportunity bits.

> It is important that the correct information about whether or not S_1 is being used is received. If the information is wrong, it is not simply a case of an error in the output data, it is a bit slip. Bit slips are generally much more serious than errors because they require, for example, subsequent multiplexers to re-align.

8.2.2 Limitations of the plesiochronous network

There are a number of features which would be desirable in a transmission network which are not supported by the plesiochronous digital hierarchy (PDH). Some of these limitations come about because of the way that the current PDH is specified. For example, transmission systems are treated as single units in the modular structure (see Fig. 8.2). The signal on the transmission link between the exchanges (optical fibre, coaxial cable etc.) does not need to comply with any standard, so both ends of a link need to come from the same supplier and opportunities for re-configuring networks by interchanging links external to an exchange are severely restricted. Furthermore, the functions of transmission systems are specified purely in terms of transporting the data from the multiplexers. In the context of a whole network, there are other supervisory and management related functions which need to be considered. For example, in the event of the failure of a link, automatic network re-configuration and automatic reporting to a maintenance centre would be desirable.

There are other limitations in the PDH which are inherent in the fact that it is plesiochronous. Consider, for example, a high capacity transmission system between two major centres (carrying, say, a 140 Mbits s^{-1}). Suppose that half way along the link there is a village requiring just one primary multiplex (2028 kbits s^{-1}) signal (Fig. 8.4). To

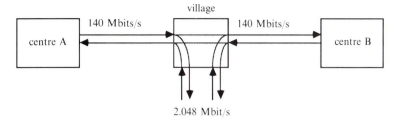

Fig. 8.4. Add & drop application.

service the village, it is necessary either to run a low capacity link in parallel with the 140 Mbits s^{-1} system, or else to access one of the 64 primary multiplexed signals carried by the 140 Mbits s^{-1} system. The latter would seem the preferable option. However, the 64 primary level signals have been asynchronously multiplexed in three stages, each involving the use of justification. The effect of the multiplexing and justification has been to scatter the 2 Mbits s^{-1} signals around the 140 Mbits s^{-1} signal. In practice, the only way to get access to the required channel is to completely demultiplex down to a primary rate signal then re-multiplex to re-construct the 140 Mbits s^{-1} signal (Fig. 8.5). The ability to 'add and drop' channels from a multiplexed signal without complete demultiplexing would clearly be useful.

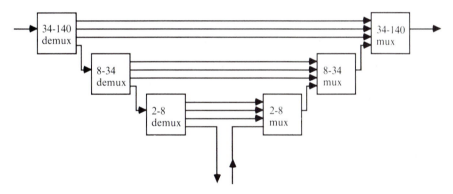

Fig. 8.5. Add & drop using plesiochronous multiplexing.

Related to 'add and drop' ('drop and insert') facilities are 'cross-connect' facilities. Fig. 8.6 shows four locations, each of which needs transmission channels to each of the others. Three pairs of high-capacity transmission systems are installed, one from each of B, C and D to A. Some channels on each of these systems need to terminate at A, while the others need to be split between the two other locations. The ratio of the splits will depend upon the capacities required. Clearly there is a need to

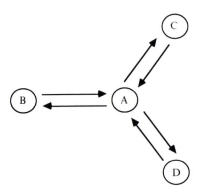

Fig. 8.6. Cross-connect.

'carve up' the capacity of each system at location A. Again, with PDH this could require complete sets of de-multiplexers and multiplexers. Access to individual channels or groups of channels, without the need for demultiplexing, would significantly simplify the equipment at location A.

'Fixes' to overcome these limitations of the PDH have been applied in various situations, but a more satisfactory solution is a complete new set of standards. The CCITT SDH and the ANSI SONET both provide the necessary standards.

8.3　The synchronous digital hierarchy (SDH)

If all the digital signals in a network derive their timing from a common source, then higher order multiplexing can be effected in the same way as primary multiplexing, without the need for justification. In principle, then, an ideal synchronous digital hierarchy, based upon the 30 channel PCM multiplex, could be constructed as in Fig. 8.7. All levels have frames similar to the primary multiplex frame and the frame period is always the same (125 μs), but higher levels have more channels per frame. Successive levels are constructed by selecting one byte from each of the contributing signals in rotation. At all levels each (64 kbits s^{-1}) channel has one byte per frame and that byte appears in the same (known) position in each frame. Thus, any single channel can be accessed from any level in the hierarchy, allowing very simple 'add and drop' facilities. Notice also that the period used by overheads (just the frame alignment word here) remains the same for successive levels.

A hierarchy as just described could be constructed for a new network which is synchronised throughout, but the reality is that the SDH needs to evolve from the pre-existing PDH, with its layers up to 140 Mbits s^{-1} as described in Section 8.2 above. Consequently CCITT recommend a

The 125 μs period originates in the pcm sampling rate of 8 kHz $(1/8 \times 10^3 = 125 \times 10^{-6})$. **At all levels in the hierarchy each channel needs a data rate of** 8×10^3 **bytes per second.**

tributary A	FAW	ch 1	ch 2	ch 3	---
tributary B	FAW	ch 1	ch 2	ch 3	---
tributary C	FAW	ch 1	ch 2	ch 3	---
tributary D	FAW	ch 1	ch 2	ch 3	---

| multiplexed signal | FAW | A ch1 | B ch1 | C ch1 | D ch1 | A ch2 | B ch2 | C ch2 | D ch2 | A ch3 | B ch3 | C ch3 | D ch3 | --- |

Fig. 8.7. Hypothetical synchronous multiplexing (FAW = frame alignment word).

network which takes as its first level a rate which can carry the 140 Mbits s^{-1} signal (Fig. 8.8). Multiplexing above this level is by synchronous byte-interleaving (as in the hypothetical system of Fig. 8.7), but below it the multiplexing and frame structure depends upon the particular tributary concerned. Initially these tributaries will mostly be plesiochronous, but as the network is gradually synchronised the tributaries can be changed to carry synchronous data. Only when all the tributaries are synchronous will the full benefits of the new network be available, but even before that point there are significant advantages in the new hierarchy.

The CCITT recommendations cover a wide range of possible tributaries, including each of the 2, 8, 34 and 140 Mbits s^{-1} plesiochronous levels, as well as the US plesiochronous levels. In fact it was one of the

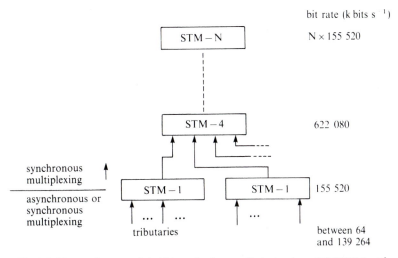

Fig. 8.8. The synchronous digital hierarchy (note: tributaries above 139 264 kbits s^{-1} can be accommodated directly into STM-N frames).

aims in drawing up the recommendation to provide a standard compatible with European and US hierarchies. The ANSI SONET standard, referred to above, is the US equivalent to SDH. SONET is discussed in Section 8.3.8.

8.3.1   STM-1

The basic building block of the CCITT SDH is the synchronous transport module level 1 (STM-1). The frame structure for STM-1 is conveniently described by a rectangle of eight-bit bytes, shown in Fig. 8.9. The transmission of the block starts with the first row, working from left to right, then moves to the second row, left to right again, and so on down to the byte in the bottom right-hand corner. Thus, although shown in a block, the overhead bytes are actually transmitted in 9 groups of 9 bytes, equally spaced throughout the frame. The frame period is 125 μs, as in the hypothetical system described above. Since there are 270×9 bytes, the data rate is $270 \times 9 \times 8 \times 8 \times 10^3 = 155\,520\,\text{kbits s}^{-1}$.

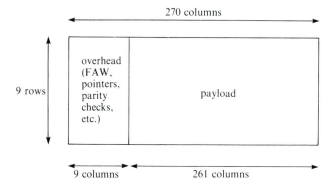

Fig. 8.9. Simplified diagram of an STM-1 frame.

The overhead (called the *Section Overhead* in the CCITT recommendation) includes the frame alignment word, parity checks, bytes to carry data and signalling channels and *Administrative Unit pointers*. The purpose of these pointers is discussed in Section 8.3.3 below.

8.3.2   Synchronous multiplexing

For level N in the hierarchy, N STM-1 signals can be synchronously multiplexed to generate an STM-N signal (Fig. 8.10), with a data rate of $N \times 155\,520\,\text{kbits s}^{-1}$.

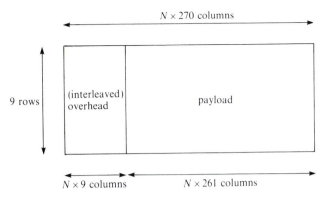

Fig. 8.10. Simplified diagram of an STM-N frame.

The procedure for multiplexing is, in principle, just byte-interleaving as in Fig. 8.7. The N 9-byte overheads become a single $9 \times N$-byte overhead.

Indirect multiplexing (e.g. $2 \times$ STM-4 or $1 \times$ STM-4 $+ 4 \times$ STM-1 to generate STM-8) is also permitted. The procedure is such as to generate an STM-N signal with the same structure as if it had been constructed directly from N STM-1 signals.

Tributary signals which require a capacity greater than can be provided by a single STM-1 signal, can be catered for by an STS-Nc signal, where the 'c' stands for 'concatenated'. This is the same as an STM-N signal, but contains a flag which indicates that the carried data should be kept together.

Aligning the frames in contributing signals is made particularly simple and efficient because of the way the data is held in the frame. This is explained in the discussion of 'pointers' below.

8.3.3 Pointers

An ingenious feature of the SDH hierarchy is the way that data is carried in the frame payload. This involves the use of pointers, and is illustrated in Fig. 8.11. One frame's worth of data consists of a block of bytes referred to as a *virtual container*. Although the virtual containers are transmitted at approximately the frame rate, they are not fixed to the frame. The position (described by a byte number) at which the virtual container starts within any one frame is given by the value of the pointer contained in the STM-1 section overhead. The advantage of this schemes is that small variations in the data rate in the container can be accommodated by allowing the virtual container to move its position in the frame. For example, if the data rate drops slightly, the virtual container starts to move forward in the frame, as illustrated in Fig. 8.12. If the data rate increases, the virtual container starts to move back. In this latter case, to accommodate the full virtual container, bytes in the overhead section of the frame are given over to the container, as shown in Fig. 8.13.

The 'pointer' is actually just a number. The number value is the location of the start of the virtual container in the payload: the first byte in the payload (top left-hand corner of the payload) is location number I. In the CCITT recommendation, the pointers are seen as being part of the payload rather than the overhead. We are treating them as part of the overhead to emphasise the fact that they are fixed in the frame, like the overhead, rather than being mobile, like the payload container.

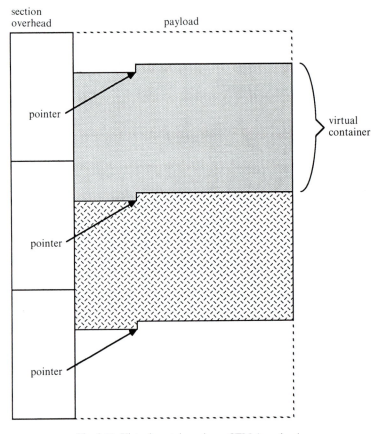

section overhead

payload

pointer

virtual container

pointer

pointer

Fig. 8.11. Virtual containers in an STM-1 payload.

This use of pointers is a variation on justification (see Section 8.2 above) to bring data into frequency synchronisation or to absorb short-term phase errors. It is used here, in preference to justification as used in the PDH, because it maintains the synchronous nature of the system insofar as examination of the frame overhead pointer bytes immediately identifies the start of the virtual container. Consequently, channels within the payload can be readily identified by using byte counters. This eases the way in which adding/dropping of channels is performed.

The pointers also simplify the alignment for multiplexing STM-1 frames to STM-N frames. N STM-1 frames will arrive at the multiplexer with random relative alignments (Fig. 8.14), whereas the outgoing STM-N has the overhead bytes grouped together in a single block. However the *payloads* of the STM-1 frames can be byte-interleaved without the need for any re-alignment, because each of the virtual containers (VC-4s or VC-3s) will have a newly calculated pointer in the STM-N overhead. The important principle is that because the virtual containers 'float' within the payloads, there is no need to align virtual

Even in a completely synchronous network, substantial phase differences or temporary frequency errors can build up in different parts of the network, due to practical limitations of the synchronisation control and transmission system jitter or wander.

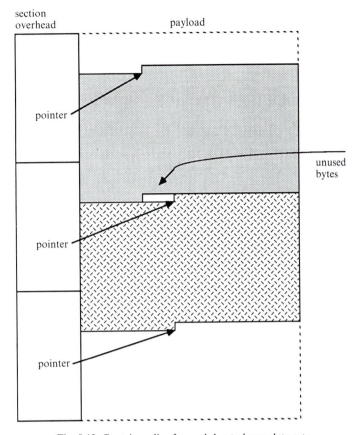

Fig. 8.12. Container slips forward due to lower data rate.

containers for multiplexing; all that is required is for the new pointer value to be calculated.

'Moving' virtual containers would mean, in practice, storing them in a buffer for a fixed period and would therefore introduce a delay. The use of pointers eliminates this delay. This is a key feature of the SDH and SONET. Without the pointer mechanism, synchronous multiplexers would require arbitrarily large storage buffers (and therefore arbitrarily large delays) to accommodate temporary variations in data rates, or else bit slips would have to be tolerated.

8.3.4 Payload structure and tributaries

It is fundamental to the philosophy of the SDH that there is a wide range of possible tributaries and corresponding range of structures used in the payload. This flexibility enables the hierarchy not only to be compatible with both the US and European plesiochronous hierarchies, but also to carry fully synchronous data as well as ATM cells.

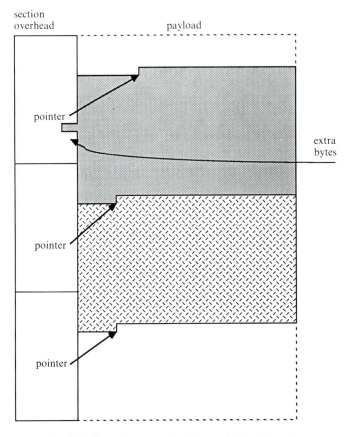

section
overhead

payload

pointer

pointer

pointer

extra
bytes

Fig. 8.13. Container moves back due to higher data rate.

Common to all tributaries is the use of pointers to align the payload to the STM-1 (or STM-N) frame. The combination of a pointer in a fixed location in the STM frame and the 'floating' virtual container to which it points, is called, in the CCITT recommendations, an *administrative unit* (*AU*). A virtual container consists of an overhead (the *path overhead* (*POH*)), and the container payload (Fig. 8.15). Although in the description of pointers above each virtual container was the size of a whole STM-1 payload, this is not the only case in which pointers are used. Individual virtual containers may be smaller than the STM-1 payload in which case there may be more than one per STM payload, with each having its own pointer. In general, however, virtual containers constitute an integer number of nine-row columns (i.e. they consist of $9 \times k$ bytes, where k is an integer) and the first column contains the path overhead.

Plesiochronous tributaries Various sizes of virtual containers are available to carry the many different plesiochronous rates. For example a

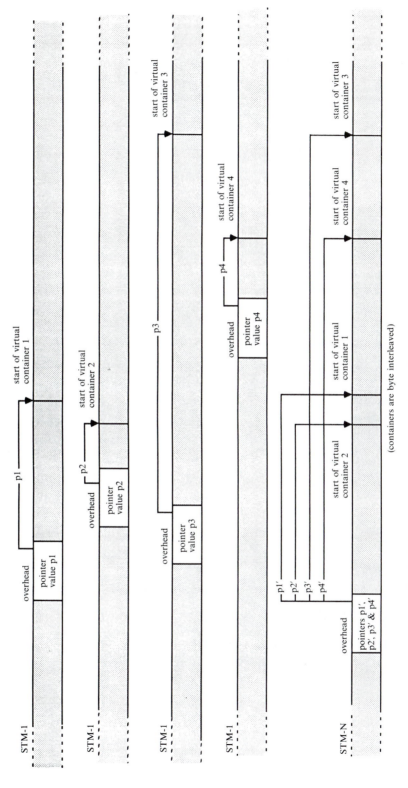

Fig. 8.14. Pointers in synchronous multiplexing. (Note: the illustration assumes that each STM-1 contains one VC-4, i.e. each STM-1 has only one higher-order virtual container.)

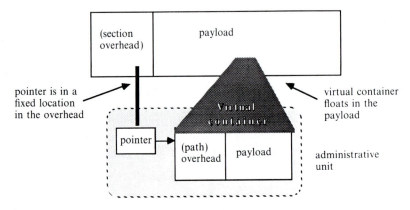

Fig. 8.15. An administrative unit.

single administrative unit per STM-1 frame, taking up the whole payload capacity, can carry one $139\,264\,\mathrm{kbits\,s^{-1}}$ channel. This type of AU is referred to as an AU-4, because the $139\,264\,\mathrm{kbit\,s^{-1}}$ signal which it can carry is level 4 in the plesiochronous hierarchy (Table 8.1). Likewise, the virtual container associated with the AU-4 is itself designated VC-4.

There also exist AU-3s, corresponding to level 3 in the plesiochronous hierarchies (Fig. 8.16). Because of their lower data rate, there is sufficient capacity in an STM-1 for four level 3 signals from the European hierarchy or three level 3 signals from the US hierarchy. The virtual containers (VC-3s) are byte-interleaved, with an administrative unit pointer associated with each virtual container to indicate the location of the start of the VC. The AU pointers are always in a fixed location of the STM-1 frame, but the VC-3s can float within the frame, as with a VC-4.

At the time of writing, the multiplexing structure has not been finalised. The current CCITT coloured book recommendations or interim documents should be consulted for up-to-date details. Individual manufacturers may also use different constructions in proprietary equipment.

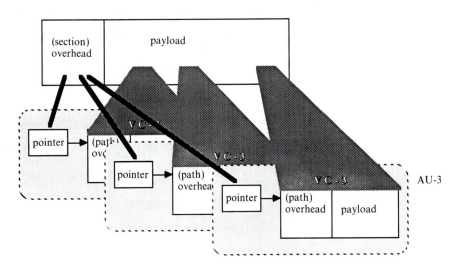

Fig. 8.16. Level 3 administrative units. Note that the virtual containers are byte-interleaved within the payload.

Signals from lower down in the plesiochronous hierarchy (levels 1 and 2 signals, from both European and US hierarchies) are also carried in virtual containers (VC-1s and VC-2s), but these containers are not put directly into the STM-1 payload; rather they are 'nested' within higher-level virtual container payloads (Fig. 8.17). At each level in the nesting, the virtual container has an associated pointer which is in a fixed position with respect to the virtual container at the next level up. A pointer plus its virtual container at levels 1 or 2 is known as a *tributary unit* (*TU*). Note that the characteristic difference between an administrative unit and a tributary unit is that an administrative unit is at the top of the hierarchy – directly in an STM frame – while a tributary unit is in a higher-level virtual container.

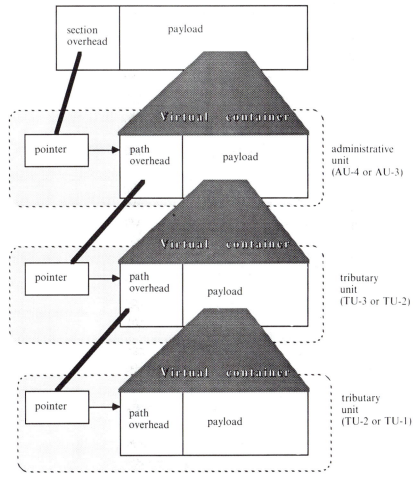

Note: An administrative unit is always in the STM frame directly, tributary units are always within higher-level containers

Fig. 8.17. Outline payload structure.

The details of the structure are rather more complicated, with several alternatives available. As mentioned above, each of the virtual container sizes can be expressed as an integer number of columns of the 270-column by 9-row STM-1 rectangle. For example, a VC-12 (which can convey one 2048 kbits s^{-1} channel) occupies four columns while a VC-11 (which can convey one 1544 kbits s^{-1} channel) occupies three columns. The dimensions of the STM-1 rectangle were, in fact, chosen so that the plesiochronous signals of both US and European hierarchies could reasonably be encompassed with a minimum of 'wastage' in terms of unused bytes.

The details of the mapping of plesiochronous data within a VC are described in CCITT recommendation G.709. It is important to note, however, that this tributary data originates from the plesiochronous network, so putting the data into the container is actually a *plesiochronous* multiplexing problem. Justification is therefore necessary to match the plesiochronous incoming data rate to the outgoing data rate of the virtual container (Fig. 8.18). The justification is carried out using justification control bits and justification opportunity bits, similar to justification in the PDH.

Since the tributary plesiochronous channel has already been constructed using justification, there is no loss of synchrony in the use of justification to map the plesiochronous channel to the VC.

In some cases level 1 and level 2 signals may derive their timing from the

> **In the numbering AU-XY, X is the plesiochronous level (1 to 4) and Y is 1 or 2, 1 indicating the lower of the frequencies at level X. A similar numbering system is used for virtual containers. Thus VC-11 carries the US primary rate (1544 kbits s^{-1}) and VC-12 carries the European primary rate (2048 kbits s^{-1}).**

> **The pointer mechanism, although able to match signals at slightly different frequencies, is *not* designed to accomplish plesiochronous multiplexing. Its use is two-fold; to absorb short-term phase and frequency differences due to fluctuations in a synchronous network, and to 'decouple' the overheads from the payload (as in aligning the section overheads for STM-1 to STM-N multiplexing).**

incoming tributary data at
data rate t bits/s, derived from the
plesiochronous network

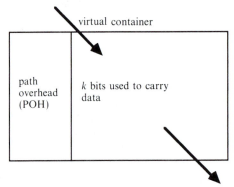

virtual container

path overhead (POH)

k bits used to carry data

Data conveyed at a rate governed by the synchronous network
(the precise rate will be $(k/19440) \times 155\ 520$ kbits/s
because there are 19440 bits in an STM-1 frame, and the
signalling rate is 155 520 kbits/s). Justification
effectively varies k to ensure that the conveyed data rate
is exactly equal to incoming data rate.

Fig. 8.18. Plesiochronous multiplexing in a virtual container.

same source as the synchronous network, in which case justification is not required and the tributaries are described as bit synchronous.

Byte synchronous payloads If a primary level signal is synchronous, an even 'better' mapping to the VC-1 is possible. This is *byte synchronous* mapping, where the primary multiplex $64\,\mathrm{kbits\,s^{-1}}$ channel bytes are transferred directly into 8 bit bytes of the containers in the synchronous hierarchy. Furthermore, the primary multiplex frame is aligned with the VC-1, such that a primary multiplex frame starts at the beginning of the VC-1 (following designated overhead bytes).

In a network based on conveying $64\,\mathrm{kbits\,s^{-1}}$ channels, this is the 'ideal' mapping for the SDH: it allows examination of pointers to locate individual $64\,\mathrm{kbits\,s^{-1}}$ channels in any STM-1 or STM-N signal.

Beware of confusion due to the separate use of the word 'synchronous' in the synchronous digital hierarchy and the synchronous transfer mode of transmission. The synchronous digital hierarchy can carry data both in synchronous transfer mode and asynchronous transfer mode.

ATM Somewhat paradoxically, synchronous transport modules are also seen as a means for conveying asynchronous transfer mode (ATM) cells in the PSTN. In an STM-1 frame, the ATM cells are transmitted serially, packaged within a VC-4 (Fig. 8.19). The virtual path and virtual channel identifiers in the cell headers distinguish between the different channels using any one link at one time. The number of cells used by each channel may be varied dynamically, according to each channel's instantaneous bandwidth requirements.

Notice the fundamental difference between multiplexed signals in synchronous transfer mode and multiplexed signals in asynchronous transfer mode: in synchronous transfer mode individual channels are identified by unique time slots in a multiplexed signal, in asynchronous transfer mode individual channels are identified by a unique number contained in the cell header. Thus in STM the capacity allocated to each channel in a multiplexed signal is fixed, whereas in ATM it can be varied.

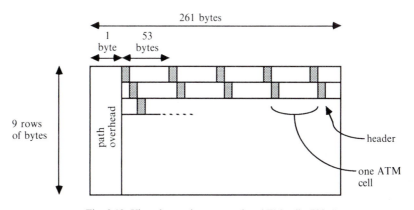

Fig. 8.19. Virtual container conveying ATM cells (VC-4).

Above the STM-1 level, use of STM-Nc frames allows individual broadband channels to operate at rates greater than the capacity of an STM-1 frame.

8.3.5   Cross-connect and add&drop

The extent to which adding and dropping is possible from STM-1 (or STM-N) frames depends upon the tributaries used by the STM payloads. Add&drop multiplex equipment can always readily locate and access virtual containers by use of the pointers, but if the data conveyed is 139 264 Mbits s^{-1} plesiochronous channels (in VC-4s), then only complete 139 264 Mbits s^{-1} channels can be handled. At the other extreme, if the payloads are structured down to VC-1s carrying byte-synchronous primary multiplex channels, then the add&drop facility can, in principle, handle individual 64 kbits s^{-1} channels.

Similarly, cross-connecting is possible at virtual container levels, or even 64 kbits s^{-1} channel level in the case of completely byte-synchronous payloads. Efficient use of a transmission network can require re-allocation of transmission capacity to cope with changing demands or equipment failure. The cross-connect facilities of SDH can be made electronically controllable, enabling a central management centre to optimise the whole network configuration. The data channels in the SDH overhead bytes will be used for telemetry in the control of the network management. Specification of the management procedures will be an important part of the final SDH recommendations.

There is obviously a parallel between switching transmission paths in cross-connect facilities and switching in a digital exchange (such as System-**X**). The distinction is that switching in an exchange is 'real-time': call-by-call. In a cross-connect switch, the routing is quasi-static, perhaps regularly changed to accommodate changing patterns of traffic, but not on a call-by-call basis. Generally, furthermore, exchange switching is customer driven, whereas cross-connect switching is usually operated as part of the network management. (Although it may be that in the future users will be provided with limited access to PTT cross-connects so that they can manage provision of loads between sites.)

8.3.6   Coding

Error detection Parity checks are used for error detection on the synchronous signals. The checks are carried out at various levels, as indicated in Fig. 8.20.

At the lowest level, there is a parity check as part of the section overhead of the STM-1 or STM-N frame which is tested and recalculated at each regenerator.

For the next level up, another parity check in the section overhead of the STM-1 or STM-N frame is only checked and recalculated at the end of a single link. In the SONET standard (see below) this part of the STM-1 or STM-N overhead, which is not needed at regenerators, is referred to as the line overhead.

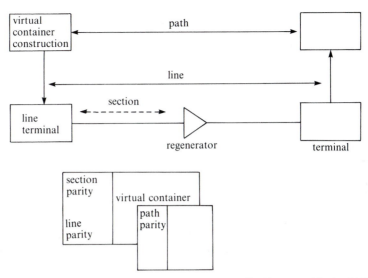

Fig. 8.20. Parity checks in an STM. (Note: The term 'line' is not used in the CCITT SDH recommendations.)

The section, line and path layers constitute a layered hierarchy similar to the hierarchy used for part 2 of this book. There is no direct correspondence between the two sets of layers, however. SONET also has a photonic layer, which does correspond to our pulse layer.

Finally, within each virtual container there is another parity check as part of the path overhead.

Each of these parity checks operates with *bit-interleaved parity* (*BIP*). A BIP-n check has n bits. The covered data (e.g. one STM-N frame) is imagined broken up into successive n-bit blocks. The first bit of the BIP is used as a parity check on the first bit of each n-bit block, the second bit is a parity check on the second bit of each n-bit block, and so on. Fig. 8.21 illustrates the method with a very simple BIP-2. Notice also that byte-interleaving N STM-1s to create an STM-N automatically turns a BIP-8 into a BIP-Nx8.

Scrambling Set–reset scrambling is used to randomise the STM-1 or STM-N signal in order to reduce the probability of long strings of data with no timing content.

Set–reset scrambling is discussed in Section 6.3. The timing content of data is discussed in Section 5.4.

The scrambler tap polynomial is $1 + x^{-6} + x^{-7}$, and setting (to all 1s) occurs immediately after the last bit of the last byte of the first row of the STM-1 or STM-N section overhead. Scrambling continues through the rest of the STM-1 or STM-N frame. The first row of the overhead is not scrambled, because this contains the frame alignment word.

8.3.7   Synchronous Optical Network: SONET

As explained above, the US equivalent of the CCITT SDH is the ANSI SONET (Synchronous Optical NETwork). SONET and SDH are (at the

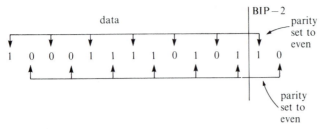

Fig. 8.21. BIP-2.

time of writing) sufficiently compatible to allow interworking between the two networks and for equipment to be constructed to be usable with either. The main difference, apart from some confusing differences in terminologies, is that SONET starts at one layer down from SDH. The basic building block of SONET is the Synchronous Transport Signal number 1 (STS-1). This has a signalling rate of $51\,840\,\mathrm{kbits\,s^{-1}}$ and is chosen so as to be able to carry one of the US plesiochronous level 3 signals (DS3, $44\,736\,\mathrm{kbits\,s^{-1}}$). The STS-1 frame structure is described by a rectangle of 90 columns and 9 rows, with the first three columns containing an overhead.

Synchronously multiplexing, by byte-interleaving, three STS-1 signals gives the STS-3, which is the same as a CCITT STM-1 signal. (Notice, for example, that the three STS-1 overhead columns interleave to become the 9 STM-1/STS-3 overhead columns.)

The SONET standards also include details at the *photonic layer*, i.e. details of the fibre characteristics and wavelengths and power levels of the light sources. The detail is sufficient, in fact, to allow 'mid-fibre meet': i.e. different manufacturers equipment on different halves of a transmission system.

8.4   Summary

In Europe, the PSTN uses a plesiochronous hierarchy for digital transmission based upon the 30-channel ($2048\,\mathrm{kbits\,s^{-1}}$) primary multiplexed signal, increasing in groups of four to $139\,264\,\mathrm{kbits\,s^{-1}}$. In the USA, a similar hierarchy is used, but based upon a 24-channel ($1544\,\mathrm{kbits\,s^{-1}}$) signal. These two hierarchies are completely incompatible. Since they are plesiochronous, both hierarchies operate by each multiplexer having its own stabilised clock source and using justification to match the incoming tributary data rates to the outgoing multiplexed signalling rate. This mode of multiplexing makes it very difficult to extract individual channels from higher order multiplexed signals for 'add&drop' facilities or for cross-connect switches.

To overcome the limitations of plesiochronous multiplexing, new transmission hierarchies based upon synchronous multiplexing have been devised. These are: SDH (synchronous digital hierarchy, described by CCITT) and SONET (synchronous optical network, described by ANSI). The two hierarchies are intended to be compatible, not only with each other but also with the existing plesiochronous networks and with the asynchronous transfer mode of transmission (ATM). This universality is made possible by 'containerisation': data is put into containers and provided these containers conform to the appropriate specifications, it does not matter what is in them.

The data rates of SDH start at the synchronous transfer module number 1 (STM-1) signal of $155\,520\,\text{kbits s}^{-1}$, which is just large enough to be able to convey a container carrying the top level plesiochronous signal. Higher rate SDH signals are constructed by synchronous byte-interleaving of numbers of tributary synchronous transfer modules to create an STM-N.

If the data source is not synchronous with the SDH signal, then 'packing' of data into containers will require justification. When containers are put into modules of the SDH (STMs), they can 'float' within the frame. Their position is tracked by pointers. Small containers are nested within larger containers. Equipment manipulating synchronous modules can locate containers within the frame by reading the pointers, and therefore can insert or extract individual containers. This facility simplifies 'add&drop' and cross-connect functions. If the original data source is also synchronous with the SDH, then the individual channels can be packed into the containers in byte-synchronous mode which allows 'add&drop' handling of individual channels, right down to the $64\,\text{kbits s}^{-1}$ level.

9

An optical fibre link

9.1 Introduction

In our study so far of the PSTN it has been tacitly assumed that digital data can be transmitted from node to node in the network at a suitably low error rate. Of course, this can only be achieved through careful engineering design, and in this final chapter we look at some aspects of such design, as exemplified by a single optical fibre link.

Fig. 9.1 shows an outline block diagram of a typical optical fibre transmission system. Although the diagram does not correspond in every detail to any specific system, the data rates shown, as well as the use of scrambling and a 5B6B line code, are widely employed in practice.

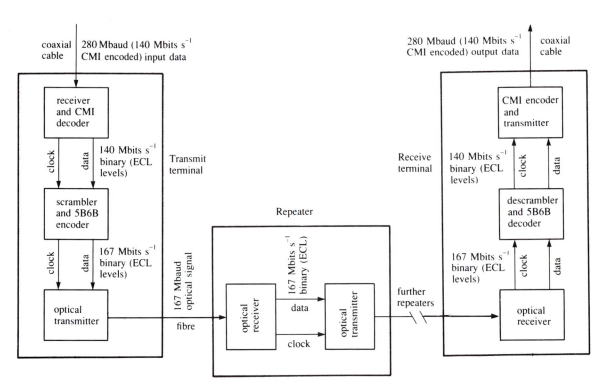

Fig. 9.1. Simplified block diagram of a typical optical fibre link.

Table 9.1. *Example of a power budget*

transmitter power	−3 dBm
receiver sensitivity	−40 dBm
System margin	37 dB
cabled fibre loss (50 km @ 0.3 dB km^{-1})	15 dB
splice loss (25 splices @ 0.5 dB)	12.5 dB
connector loss (3 connectors @ 1.0 dB)	3 dB
repair margin (10 splices)	5 dB
Route losses	35.5 dB
Excess margin	1.5 dB

The major system elements are:

a transmit terminal

a receive terminal

repeaters

the fibre itself

A single direction has been shown in Fig. 9.1, but in practice two sets of equipment are usually combined to provide a bi-directional link, with a separate fibre used for each direction.

A fundamental parameter of any digital transmission system is the distance which can be achieved between repeaters. This is determined by: the transmitted signal power; the route losses (especially the attenuation per kilometre of the transmission medium); and the sensitivity of the receiver (required signal level for a given error rate). In a particular system these quantities might be analysed by means of a power budget, such as Table 9.1. Note that by working in decibels the various contributions can simply be summed.

For an optical fibre system the light source at the transmitter is usually an infra-red light emitting diode (LED) or laser diode. Simple on–off pulses are generally transmitted. Maximum transmitted power is limited by manufacturing technology and safety considerations: the power from a laser diode can be sufficient to damage eyesight, even after transmission over several kilometres of fibre. The choice between LED and laser diode for a given application will be influenced by various factors, including cost.

For a required transmitter power towards the upper end of the LED range it will often be more cost effective to select a relatively low-power laser diode rather than a high-performance LED.

The main element in the route losses for a trunk optical fibre system will be the attenuation of the fibre. It is a remarkable achievement of the fibre manufacturers that the fibre currently in use has an attenuation close to the theoretical minimum for the optical wavelengths used. This is around 0.2 dB for a wavelength of 1300 nm and slightly less for a wavelength of 1500 nm. Because of this low attenuation, very long distances between repeaters are possible, currently up to a few hundred kilometres. In fact, many transmission systems do not need any repeaters, since the total route length is possible in a single 'hop' from the transmit terminal to the receive terminal. For short transmission links the major route losses may not be the fibre attenuation. Losses in connectors, for example, may be more significant.

In Chapter 3 it was shown that the error probability for a unipolar signal degraded by additive white Gaussian noise was given by $P_e = Q\{(S/N)_v/2\}$, where $(S/N)_v$ is the signal to noise (voltage) ratio at the regenerator at the sampling instant. In Chapter 4, the analysis was generalised to an optimal detector (matched filter), for which the theoretical error rate (assuming AWGN) is determined solely by the ratio between the received symbol energy and the noise power spectral density. This would seem to imply that for a given received pulse energy the sensitivity of a receiver is determined solely by the received noise power spectral density. This is not so for a real receiver, however, for two reasons – one theoretical and one practical:

1. Even in an ideal optical receiver, the dominant noise would not be AWGN, so the quantitative results of Chapters 3 and 4 do not hold. In fact, the dominant noise would be a form of shot noise inherent in the way photons interact with matter. Photon arrival can be closely modelled by the *Poisson distribution*: If the mean number of photons received in a symbol period representing a '1' is μ, then the probability $P(n)$ of n arriving in any particular data '1' period is given by:

$$P(n) = \mu^n e^{-\mu}/n!$$

For high photon arrival rates, the Poisson distribution can be approximated by the Gaussian distribution introduced in Chapter 3. For low light intensities, though, the Poisson model is needed to give quantitative estimates.

2. Current practical receivers introduce additional noise which reduces the sensitivity, compared with an ideal receiver, by several orders of magnitude.

This second point makes it worthwhile putting a lot of design effort into getting the best possible receiver performance. This chapter shows how the signal and system models of Chapter 2 and the noise models of

Chapter 3 can be applied to this problem, and also considers how some of the theoretical ideas about pulse transmission introduced in Chapter 4 are reflected in practical engineering.

9.2 Receiver signal to noise ratio

Although quantitative estimates of error rate cannot be derived from S/N in the way described in Chapter 3, this parameter is still very important: a high signal to noise ratio will indicate a low error rate and vice versa. Consequently, by deriving a formula for the signal to noise ratio, it is possible to arrive at various conclusions about the receiver design and the relative importance of the different sources of noise.

9.2.1 The receiver model

Fig. 9.2 shows a general model of an optical receiver. The received light is converted to an electrical signal by a photodiode. This is a semiconductor device which, when suitably biased, provides a current out proportional to the incident light. In an ideal photodiode each incident photon would contribute one electron to the current. In a practical photodiode some photons fail to add an electron to the current, so the ratio of electrons to photons (which is less than 1) is used as a parameter of the device and is known as the quantum efficiency, η. An alternative measure of the quantum efficiency of a photodiode is its responsivity, R. The responsivity is the ratio of the current out from the diode to the light power incident upon it.

A type of photodiode commonly used in telecommunication systems is known as a *PIN diode*. PIN stands for positive-intrinsic-negative and

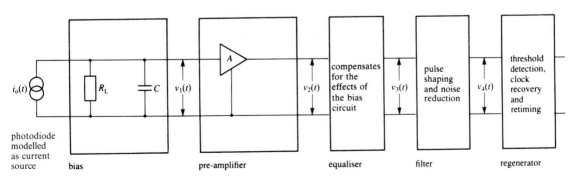

Fig. 9.2. A general optical receiver model.

refers to the physical construction of the device; further information can be found in, for example, Jones. An alternative type of photodiode sometimes used in optical fibre receivers is known as an *avalanche photodiode (APD)*. An APD is constructed and used in such a way that each electron released by a photon is accelerated in an electric field and collides with other electrons, causing an 'avalanche' of electrons. The effect is to increase the output current for a given received power. This is modelled by an avalanche gain, M, so the current out from an APD is given by MRP, where P is the incident light power. In this section we shall assume that an APD is used.

The current from the photodiode is passed through a resistor, R_L, to produce a voltage $v_1(t)$ which is then amplified by the pre-amplifier. The capacitor, C, shown in Fig. 9.2 is included to model the capacitance of the photodiode, the amplifier input and the circuit wiring. As will emerge from the analysis which follows, this capacitance has a very significant effect upon the receiver performance. The combination of R_L and C forms a low-pass filter with a 3 dB cut-off frequency $1/2\pi R_L C$. This low-pass filtering effect needs to be compensated for by the equaliser shown immediately following the pre-amplifier. Note that the equaliser compensates for *receiver*, rather than *channel* characteristics.

The final block (before regeneration) shown in Fig. 9.2 is the pulse shaping filter. This is designed to minimise intersymbol interference (ISI) and at the same time remove noise outside the frequency range of interest.

In order to analyse this system it is necessary to make a number of modelling assumptions. The equaliser will be assumed to compensate perfectly for the effect of the input bias RC-section over the frequency range of interest. The pulse-shaping filter will be modelled simply as an ideal low-pass (brick-wall) filter, with cut-off frequency B equal to half the signalling rate. Finally, despite the comments above, the various independent sources will be treated as though they are all additive, white and Gaussian.

Because of these simplifying assumptions, it will not be possible to make accurate numerical estimates of error probabilities. However, it will be possible to decide which of the various noise sources are dominant under given circumstances, and use this information to make sensible decisions about the trade-offs inherent in practical receiver design. We shall give the receiver noise analysis in some detail, but the most important part of this Section is 9.2.4, where consequences for design are discussed.

A brick-wall frequency response is very different from that of any filter which would be used in a practical system, and would be inadequate as a model if we were concerned with an accurate quantitative assessment of the receiver performance – especially intersymbol interference. However, the salient feature of the filter for the *noise* analysis is that its bandwidth is proportional to the signalling rate. The brick-wall response has this feature, and is quite adequate for the present purposes.

9.2.2 Modelling the signal power

Consider first how to calculate the signal voltage $v_3(t)$ at the output of the equaliser of Fig. 9.2. The APD is modelled as an ideal light-controlled current source so that the current i_0 is given by

$$i_0(t) = MRp_i(t)$$

where M is the avalanche gain, R is the device responsivity, and $p_i(t)$ is the incident optical power.

From the above modelling assumptions the overall frequency response up to and including equalisation is completely flat, so the voltage $v_3(t)$ is directly proportional to the current $i_0(t)$ (and thus, by implication, to the incident optical power):

$$v_3(t) = AR_L i_0(t)$$

where A is the voltage gain of the amplifier.

Neglecting dispersion in the fibre and assuming that on–off keying is used, the optical signal, and hence the equalised voltage signal $v_3(t)$, will approximate to a rectangular pulse stream. The binary 1s will correspond to a constant incident optical power p_1 and 0s to zero incident power. When the incident light represents a binary one, then

$$i_0(t) = MRp_1 = \text{constant}$$

and

$$v_{3(1)} = AR_L MRp_1$$
$$= AR_L Mi_1$$

where $i_1 = Rp_1$ is the (constant) photocurrent generated by the (constant) optical power representing a binary 1.

To find the filtered signal level $v_4(t)$ at the sampling instant we need in theory to calculate the effect of the filter on the rectangular pulse train $v_3(t)$. The approximate effect of the brick-wall filter on the rectangular pulse train is shown in Fig. 9.3. We shall therefore assume that the value of $v_4(t)$ at the sampling instant – the voltage difference between the '1' and the '0' level – is still equal to $v_{3(1)}$. To the accuracy required for this analysis, it is a reasonable approximation.

Assuming the usual dissipation in a 1 ohm resistor, the signal power for a binary 1 prior to threshold detection is therefore

$$S = v_{3(1)}{}^2 = A^2 R_L{}^2 M^2 i_1{}^2$$

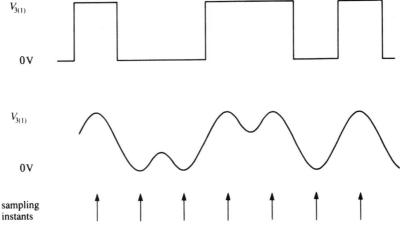

Fig. 9.3. Approximate effect of brick-wall filter on pulse train.

9.2.3   Noise sources

Fig. 9.4 is a model of the receiver showing the various contributions to the noise, which will now be considered in turn. (Note that the various noise sources are modelled as spectral densities in mean-square current or voltage form.)

Shot noise This is modelled as in Chapter 3 by a current source with a mean-square current spectral density (single sided) of

$$\{i_s^2\} = 2eI_t$$

where I_t is the current flowing through the photodiode and e is the charge on an electron (1.6×10^{-19} coulombs). There are two components to I_t. There is the signal current due to the received optical signal, but there is also a 'leakage' current, known as dark current I_d which flows through the device even in the absence of an incident optical signal. This model of shot noise predicts a noise current density which will be higher during data 1s than 0s – because of the signal current during 1s. For simplicity, this complication will be ignored and the value equivalent to a data 1 will be used all the time. This will give a pessimistic estimate of the noise.

In an avalanche photodiode (APD), the shot noise (like the photocurrent itself) will be multiplied, giving a multiplied shot noise density of

$$M^2\{i_s^2\} = 2M^2eI_t$$

Because the avalanche gain itself involves a random process (the random collisions between electrons) there is a *gain randomness* which is an

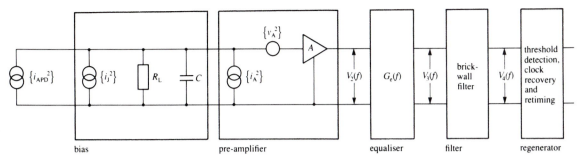

Fig. 9.4. Noise model of the receiver.

additional source of noise in an APD. It is found that a reasonable way of modelling this in practice is to replace M^2 by the term M^{2+x}, where x is a constant, typically around 0.3 for diodes constructed from silicon and between 0.7 and 1.0 for germanium and GaAs materials.

The total noise from the APD is therefore modelled by the current squared spectral density shown in Fig. 9.4:

$$\{i_{APD}{}^2\} = 2M^{2+x}eI_t$$

Thermal noise from the load resistor This is modelled, again as in Chapter 3, by an additional current source (in parallel with the APD noise source):

$$\{i_J{}^2\} = 4kT/R_L$$

where k is Boltzmann's constant $1.381 \times 10^{-23}\,\mathrm{JK^{-1}}$

The amplifier The noise introduced by the amplifier will depend upon both the design and the individual components. A useful general model is shown as the pre-amplifier block of Fig. 9.4, in which noise from various amplifier components is modelled by an equivalent *current* noise source, with mean-square spectral density $\{i_A{}^2\}$, together with a *voltage* noise source, with spectral density $\{v_A{}^2\}$. These two sources are assumed to be independent, additive, white and Gaussian – this is a reasonable approximation which makes the analysis tractable.

Fig. 9.4 can be tidied-up somewhat by combining the three current noise terms to give Fig. 9.5.

The first step in calculating the total noise power is to calculate the total noise *voltage* spectral density at the input to the amplifier, taking into account the frequency response of the bias section. This spectral density will be made up of two terms: the amplifier voltage noise source itself, plus a *voltage* spectral density generated across the bias circuit by the three *current* noise sources.

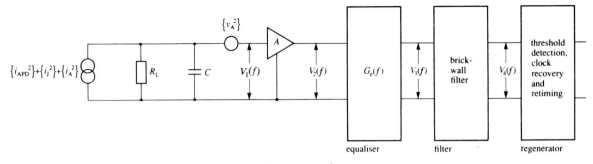

Fig. 9.5. The noise model redrawn to combine noise current sources.

Circuit analysis gives the frequency response (complex impedance) of the RC bias section as

$$H(f) = V_1(f)/I_0(f) = R_L/(1 + j2\pi f R_L C)$$

which we shall call Z_b.

The bias section output noise *voltage* density is hence

$$(\{i_{APD}{}^2\} + \{i_J{}^2\} + \{i_A{}^2\})|Z_B|^2$$

Adding this to the amplifier *voltage noise* term gives a total noise voltage density at the input of the amplifier of

Here we apply the general rule for linear systems: output power spectrum = input power spectrum $\times |H(f)|^2$.

$$\{v_{n1}{}^2\} = (\{i_{APD}{}^2\} + \{i_J{}^2\} + \{i_A{}^2\})|Z_B|^2 + \{v_A{}^2\}$$

To complete the noise analysis, a similar step needs to be carried out for the equaliser and filter, in order to determine the spectral characteristics of the noise at the input to the regenerator. An ideal equaliser would have a frequency characteristic which perfectly compensates for that of the bias circuit

$$G_e(f) = V_3(f)/V_2(f) = 1 + j2\pi f R_L C = R_L/Z_B$$

The overall frequency response of amplifier, equaliser and filter combined is thus

This expression is dimensionless since it relates two voltages. Note also that the overall frequency response up to and including equalisation is $Z_B \times A \times R_L/Z_B = AR_L$, the factor used earlier to relate v_3 and i_0.

$$V_4(f)/V_1(f) = AG_e(f)G_s(f) = A(R_L/Z_B)G_s(f)$$

The pulse shaping brick-wall frequency response $G_s(f)$ has the effect of making the spectral density zero for frequencies greater than B and leaving it unaltered for frequencies less than B.

Hence

$$V_4(f)/V_1(f) = A(R_L/Z_B) \quad \text{for } f < B$$
$$= 0 \quad \text{for } f > B$$

The power spectrum relationship for noise voltage densities then becomes

$$\{v_{n4}{}^2\} = \{v_{n1}{}^2\}A^2R_L{}^2/|Z_B|^2$$
$$= [(\{_{APD}{}^2\} + \{i_J{}^2\} + \{i_A{}^2\}).|Z_B|^2 + \{v_A{}^2\}]A^2R_L{}^2/|Z_B|^2$$
$$\text{for } f < B,$$

With some tidying-up, and substituting for $|Z_B|^2$ we have

$$\{v_{n4}{}^2\} = (\{i_{APD}{}^2\} + \{i_J{}^2\} + \{i_A{}^2\})A^2R_L{}^2 + \{v_A{}^2\}A^2(1 + 4\pi^2f^2R_L{}^2C^2)$$

over the range of interest.

The total noise power (mean-square voltage) is given by integrating this voltage spectral density with respect to frequency over the bandwidth of the filter (0 to B). This leads to the final expression for the noise power:

$$N = (\{i_{APD}{}^2\} + \{i_J{}^2\} + \{i_A{}^2\})A^2R_L{}^2B + \{v_A{}^2\}A^2(B + \tfrac{4}{3}\pi^2B^3R_L{}^2C^2)$$

Combining this expression with the earlier one for the signal power leads to the signal to noise (power) ratio

$$S/N = \frac{A^2R_L{}^2M^2I_1{}^2}{(\{i_{APD}{}^2\} + \{i_J{}^2\} + \{i_A{}^2\})A^2R_L{}^2B + \{v_A{}^2\}A^2(B + \tfrac{4}{3}\pi^2B^3R_L{}^2C^2)}$$

9.2.4 Consequences for receiver design

Expanding $\{i_{APD}{}^2\}$ and $\{i_J{}^2\}$ and re-arranging the final expression for signal to noise ratio leads to:

$$S/N = \frac{I_1{}^2}{2M^xeI_1B + \dfrac{4kTB}{M^2R_L} + \dfrac{\{i_A{}^2\}B}{M^2} + \dfrac{\{v_A{}^2\}B}{M^2R_L{}^2} + \dfrac{4\pi^2B^3C^2\{v_A{}^2\}}{3M^2}}$$

$$\qquad\qquad\text{(a)}\qquad\text{(b)}\qquad\quad\text{(c)}\qquad\text{(d)}\qquad\qquad\text{(e)}$$

shot	thermal	amplifier	amplifier	
noise	noise	current	voltage	
		noise	noise	

(9.1)

The numerator of Equation (9.1) is determined by the received signal power, and will be subject to practical limitations on its maximum value. To maximise the S/N, and hence minimise error rate, we must seek to minimise the denominator of the expression. The form of Equation (9.1) allows a number of useful conclusions to be drawn.

Optimum value of avalanche gain Increasing the avalanche gain, M, decreases the noise terms (b), (c), (d), and (e) but increases the noise term (a). It follows from this that with all else fixed (signalling rate, dark current, pulse shaping, pre-amplifier noise, R_L and C), there is an optimum value of M which minimises the denominator, and hence maximises S/N.

This (theoretical) optimum value can be determined from Equation 9.1 by using calculus to determine the minimum value of the denominator with M as the variable. However, this procedure leads to a rather cumbersome expression, and it is sufficient to look at general features. In particular, note that the B^3 term (e) is decreased by increasing M. This means that in general, higher signalling rate systems require higher values of M.

Similarly, it might be thought that M should always be greater than 1: with $M = 1$ the sum of terms (b) to (e) will always be greater than term (a) for real components. This implies that an APD will always be better than a PIN diode, which has no avalanche gain. In practice, however, the dark current for an APD will generally be greater than for a PIN diode, and the more complex bias circuitry required by an APD leads to other differences in circuit parameters. The result is that for lower signalling rate systems PIN diodes give the better performance, but that the B^3 terms leads to APDs giving the better performance at higher signalling rates. The cross-over frequency depends upon the individual devices used and the circuit designs. It can be as high as several hundred megabaud – so the reference to 'lower' signalling rates above was very much relative!

Choice of the load resistor Both terms (b) and (d) in Equation 9.1 are decreased by increasing R_L. There is a practical limit to the size of R_L which may be used, however, because increasing R_L leads to the need for more high-frequency gain in the equaliser. It is difficult to design amplifiers with high gain at high frequencies; furthermore, attempting to achieve this results in *noisier* amplifiers (larger $\{v_A^2\}$ and $\{i_A^2\}$) which counteracts the benefits of the larger R_L!

A practical design aims to use a value of R_L which is just sufficient to prevent (b) or (d) dominating as noise sources. A typical value might be in the range $100\,\text{k}\Omega$–$10\,\text{M}\Omega$.

The shot noise limit Most of the noise terms in Equation 9.1 can, theoretically, be reduced indefinitely (none, of course can be eliminated entirely in practice) by increasing R_L, using better amplifiers with lower noise or by increasing M: all of them, in fact, except for the shot noise term (a). This could be reduced to $2eI_1B$ by making $M = 1$, but I_1 and B cannot be reduced without affecting the signal power. This term therefore represents a fundamental limit known as the *shot noise limit*.

The shot noise limit is a consequence of the quantum nature of light and charge carriers. Quantum fluctuations set an absolute lower limit to the mean photon arrival rate for a given error probability, since on average several photons per signalling interval need to be transmitted to give a high probability of at least one arriving at the appropriate time. Analysis assuming Poisson statistics shows that for on-off signalling an average of 20 photons per data '1' must be received in order for the error rate due to fluctuations in photon arrival to be less than 1 in 10^9. (See, for example, Jones for the derivation of this result.)

A practical way of making the shot noise term the dominating noise, is to increase M sufficiently for (a) to be greater than (b) + (c) + (d) + (e). A receiver designed to operate in this way is referred to as 'shot noise limited' and has a signal to noise ratio from Equation 9.1 of

$$S/N = I_1^2/(2M^x e I_t B) \approx I_1/(2M^x e B)$$

where the latter approximation applies if the photodiode dark current is small ($I_d \ll I_1$).

Shot noise was modelled in the above analysis by its continuous data 1 value. Because of this approximation, and because photon arrival is a Poisson, rather than a Gaussian process, the S/N formula above cannot be used in any specific error rate calculations using the Q function based on a Gaussian distribution. However, the formula *can* be used to estimate the order of magnitude of shot noise limited performance.

The high frequency limit Because of the B^3 factor, term (e) dominates in the limit of high signalling rate systems. The signal to noise ratio in this case is:

$$S/N = 3M^2 I_1^2/(4\pi^2 B^3 C^2 \{v_A^2\})$$

Apart from M, which has already been discussed, the elements of significance to the design are C^2 and $\{v_A^2\}$. It is clearly important to minimise both.

A common approach to achieving this aim is to use a field effect transistor (FET) for the first stage of the pre-amplifier, which provides a low $\{v_A^2\}$ and also has a high input impedance to allow a high R_L. To minimise C, the PIN diode and pre-amplifier are built on a 'hybrid circuit'. A 'hybrid' is a ceramic plate on which electronic components are assembled with a minimum of packaging and with the shortest possible interconnections. Such an arrangement (referred to as a *PINFET hybrid*) allows C to be kept small, since a significant contribution to C can come from the capacitance of the wiring between components.

Typical values from this construction are

$$C \approx 1\,\text{pF}$$
$$\{v_A^2\} \approx 10^{-17}\,\text{V}^2\,\text{Hz}^{-1}$$

EXAMPLE

Compare the relative magnitudes of each of the noise sources in a PINFET receiver with the following parameters

$R_L = 1\,M\Omega$, $C = 1\,pF$, $\{i_A{}^2\} = 10^{-28}\,A^2\,Hz^{-1}$, Temperature $= 300\,K$
$\{v_A{}^2\} = 1.6 \times 10^{-17}\,V^2\,Hz^{-1}$, $B = 90\,MHz$, $I_1 = 50\,nA$, $I_d = 0.1\,nA$

(Boltzmann's constant $k = 1.381 \times 10^{-23}\,JK^{-1}$)
Which is the dominant term?
Calculate the resulting signal to noise power ratio.

Solution Calculating each of the terms (a) to (e) in Equation 9.1, with M, the avalanche gain $= 1$ because the system is using a PIN diode, gives:

(a) Shot noise:

$$2eI_t B = 1.44 \times 10^{-18}\,A^2$$

(b) Thermal noise:

$$4KTB/R_L = 1.49 \times 10^{-18}\,A^2$$

(c) Amplifier current noise:

$$\{i_A{}^2\}B = 9 \times 10^{-21}\,A^2$$

(d) Amplifier voltage noise:

$$\{v_A{}^2\}B/R_L{}^2 = 1.44 \times 10^{-21}\,A^2$$

(e) Amplifier voltage noise:

$$4\pi^2 B^3 \{v_A{}^2\} C^2/3 = 1.53 \times 10^{-16}\,A^2$$

(e) is the dominant term – being two orders of magnitude larger than any of the others, so the approximate S/N is given by

$$S/N = I_1{}^2/(a + b + c + d + e) \approx I_1{}^2/(e)$$
$$= (50 \times 10^{-9})^2/(1.53 \times 10^{-16})$$
$$= 16 \; (12\,dB) \; (2 \text{ significant figures})$$

EXERCISE 9.1

The PIN is replaced by an APD, so that C becomes 4pF as a result of the APD/pre-amplifier circuit design. All other components have the same values. The dark current of the APD is $5\,nA$ and APD gain randomness is modelled as described above with $x = 0.7$.

What value of avalanche gain would make the shot noise equal to the largest of the other noise terms? Show that the new signal to noise ratio at this gain is approximately 21 dB.

9.3   Pulse shaping

The last section assumed the pulse shaping filter to be a brick-wall filter with cut-off frequency equal to half the signalling rate. This was for convenience in noise modelling; such a filter would be neither realisable nor desirable. For a more detailed analysis using a realistic pulse shaping filter, the brick-wall model would need to be replaced by some more appropriate transfer function $G_s(f)$. Calculations such as these are performed during the design of a receiver with the aid of computer programs. Some simple observations are possible without detailed calculations, however. A useful starting point is to recall that if the noise before the pulse shaping had been additive white Gaussian noise, then the optimum solution (for maximum signal to noise ratio) would have been a matched filter. There are several departures from additive white Gaussian noise here. For example, because of the colouring of the noise by the equaliser, there is a B^3 noise term in the expression for signal to noise ratio which leads to a disproportionate contribution to the total noise power from higher frequency noise. It is therefore beneficial to reduce the high-frequency gain at the receiver, provided the loss of high frequency in the signal is compensated for by using a transmitted pulse shape which has enhanced high-frequency components. In theory, an ideal transmitted pulse shape can be determined. In practice, half-width rectangular pulses are commonly used because they are easy to generate; the complex circuitry required to generate any other pulse shape is unlikely to be worthwhile.

> A similar effect was noted in Chapter 4 in the context of equalisation for copper line transmission. Reducing the width of the pulses increases the high-frequency content of the spectrum, as discussed in Chapter 2.

Consideration of intersymbol interference (ISI) is of utmost importance. The starting point is likely to be a raised cosine shaping for the overall filtering. Some modification may be employed to compromise between maximum signal to noise ratio and minimum intersymbol interference.

9.4 The integrating effect of the high-impedance receiver

The receiver analysed in Section 9.3, in which the load resistor R_L is kept high to minimise noise, is an example of a design known as a 'high-impedance' receiver. Apart from the difficulty of providing the necessary high-frequency gain in the equaliser, a high-impedance receiver suffers from another problem arising from the value of the bias circuit cut-off frequency in comparison with the signalling rate. For example, with the figures used in the example of Section 9.3.4, the cut-off frequency of the bias circuit is

$$1/2\pi R_L C = 1/2\pi \times 10^6 \times 10^{-12} = 159\,\text{kHz}$$

The signals conveyed by the system contain components well above this (if B is the Nyquist frequency, then the signalling rate in the example is

180 Mbaud). The equaliser will compensate for the filtering, but consider for the moment the effect of the low-pass bias section on such signals *prior* to the equaliser.

The effect of a low-pass filter on signals with frequency components well above its cut-off is similar to the mathematical operation of integrating the signal over time. From Chapter 2, a first-order low-pass filter has a frequency response function of the form $k/(1+jf/f_o)$ where f_o is the 3 dB cut-off. For $f \gg f_o$ this becomes approximately $k/(jf/f_o)$, the frequency response of an integrator. Because of this integrating effect of the low-pass filter at the input, a high-impedance receiver is sometimes known as an 'integrating' receiver.

Alternatively, the effect can be interpreted in the time-domain. If an input pulse is much shorter in duration than the filter time constant (i.e. most frequency components are much higher than the cut-off), only the early part of the exponential rise of the pulse response can be seen.

The pulse response therefore approximates to a ramp, the characteristic response of integration of a square pulse with respect to time. The effect on a digital signal of this 'integrating' property is illustrated in Fig. 9.6(a). Isolated input pulses cause small 'ripples' on the output, while longer strings of consecutive '1's and '0's cause approximate ramps. The pre-amplifier must ensure that the signal is well above the rms value of subsequently introduced noise so, after amplification, the ripples – which contain the signal information – must be much larger than the noise. But amplifying the ripples amplifies the ramps equally. So having set an amplifier gain sufficient to raise the ripples above the noise voltages, it is necessary to ensure that the ramps do not cause saturation, otherwise signal information can be lost, as shown in Fig. 9.6(b).

The significant factor is the ratio of the height of the ripples to the extremes of the voltage reached. This ratio is, in fact, the Digital Sum Variation (DSV) of the line code used on the system. From Chapter 5, the running digital sum of transmitted data is determined by counting $+1$ for a positive pulse and -1 for a negative pulse and keeping a running sum of this for each data symbol transmitted. So integrating, as in Fig. 9.5, is

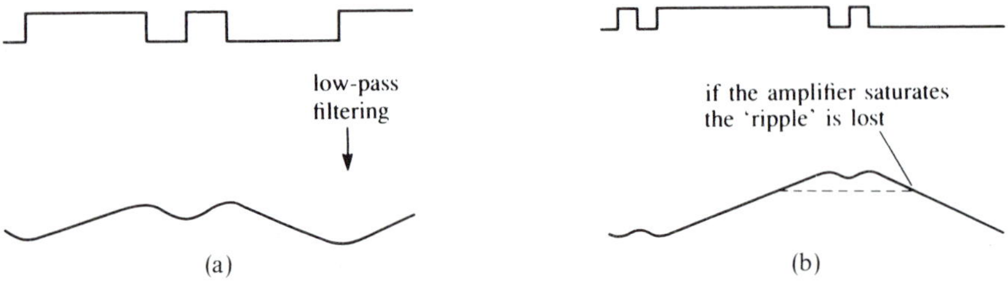

> **The pulse response of a low-pass filter shown earlier as Fig. 2.3 was for a pulse whose duration was similar in magnitude to the filter time constant.**

Fig. 9.6. (a) Integrating effect of a low-pass filter. (b) Effect of amplifier saturating.

low-pass
filtering

(a)

if the amplifier saturates
the 'ripple' is lost

(b)

Fig. 9.7. *RC* first-order low-pass (a) and high-pass (b) filters.

doing effectively the same thing: the output moves up some voltage (say V) for each '$+$' and down the same amount for each '$-$'.

If the height of a ripple is given by V, then the difference between the positive and negative extremes will be given by the code DSV times V. The *ratio* of the height of the ripples to the extremes, is therefore just given by the DSV.

The argument is parallel to the argument relating baseline wander to the code DSV in Chapter 5. Compare a simple *RC* low-pass filter with an *RC* high-pass filter (Fig. 9.7). In both cases the voltage across the capacitor is proportional to the running digital sum. In the case of the low-pass filter, the output voltage *is* this voltage proportional to the running digital sum. In the case of the high-pass filter, the output voltage contains the input voltage *offset* by the voltage proportional to the running digital sum.

The practical consequence of this is that using a high impedance receiver design means that the code used must have a restricted DSV, and that it is desirable to keep the DSV as low as possible.

9.5 The trans-impedance receiver

Limitations of the high-impedance receiver can be overcome to some extent by the use of feedback around the pre-amplifier.

The configuration used (Fig. 9.8) is known as a *trans-impedance amplifier*, because it is designed to give a voltage out proportional to the current in – it has transfer characteristics which have the dimensions of ohms. It is used in a *trans-impedance receiver* in place of the load resistor R_L and the pre-amplifier of the high-impedance design, leading to a receiver configuration as in Fig. 9.9.

Applying Kirchhoff's laws to derive the input impedance of a trans-impedance amplifier leads to the expression R_f/A, where A is the amplifier

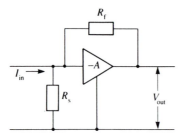

Fig. 9.8. A trans-impedance amplifier.

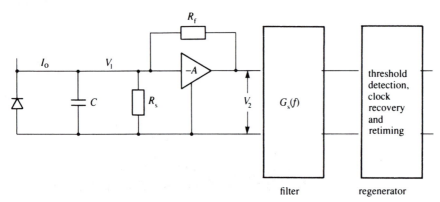

Fig. 9.9. Trans-impedance receiver configuration.

gain and R_f the feedback resistor as in Figs. 9.8 and 9.9. The analysis assumes that $A \gg 1$ and $AR_f \gg R_s$ as would be the case in a practical system.

Using this value for the input impedance gives a bias circuit transfer characteristic (frequency response):

$$Z_1 = V_1(f)/I_0(f) = (R_f/A)/(1 + j2\pi f R_f C/A) = R_f/(A + j2\pi f R_f C)$$
$$\approx R_f/A, \text{ provided that } R_f/A \gg 1/2\pi f C$$

The noise equivalent circuit is as shown in Fig. 9.10. The important point to notice is that although the input impedance is R_f/A, the associated thermal noise is that of a resistor *equal to the parallel combination of R_f and R_s*. The amplifier voltage noise is also changed. A full analysis of the trans-impedance receiver leads to an expression for signal to noise ratio identical to that of a high-impedance receiver (i.e. Equation 9.1) but again with the parallel combination of R_s and R_f taking the place of R_L.

To get the benefits of a large R_L it is therefore necessary to ensure that both R_f and R_s are large. Provided that A is large it is possible to do this

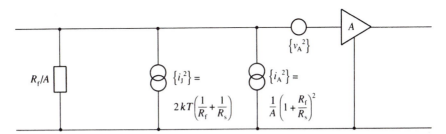

Fig. 9.10. Noise model of a trans-impedance receiver.

while keeping R_f/A small, and hence the input impedance low. So it is now possible to arrange for the low-pass filtering of the bias circuit to have a cut-off frequency well above the frequencies of interest, while not compromising the receiver noise performance. An equaliser is no longer needed; neither does the line code now need a restricted DSV, because the bias circuit no longer tends to integrate long strings of data 1s.

The trans-impedance receiver seems from the above description to offer the ideal solution for a receiver design. It is a commonly used configuration for optical receivers, but there are other considerations which mean that it is not always ideal. The analysis outlined above is still based upon the assumption of ideal components. In practice, because of the use of feedback in the trans-impedance design, there can be problems of stability due to phase shifts around the feedback loop.

It turns out that with real components, and using a code with restricted DSV, the best sensitivity is usually obtained from a high-impedance receiver, and these are used in many overland systems in Europe. For some systems, though, the absence of the need for a restricted DSV in the line code is so attractive that a trans-impedance design is the best compromise: examples include the TAT-8 transatlantic optical fibre transmission link and SONET-compatible systems in the USA.

9.6 Summary

Receiver design is a critical aspect of the error performance of an optical fibre link. Error rate is determined by the signal to noise ratio at the threshold detector at the sampling instant. A mathematical model of an optical receiver can be derived, leading to an expression for this S/N ratio. Because of the modelling assumptions made, however, the expression for S/N is not sufficiently accurate for numerical predictions of error rate to be made; however, it does allow a number of conclusions to be drawn regarding the optimisation of receiver performance. Some of these are:

the receiver load resistance should be as large as practicable for
  high-impedance designs

for good high-frequency performance, it is vital to minimise the
  capacitance of the photodiode junction, the pre-amplifier input,
  and the wiring between the two

when using an APD there is an optimum value of avalanche gain
  giving a maximum S/N

the optimum avalanche gain is higher for systems operating at
  higher signalling rates

For receivers with a high load resistance (integrating receivers) the
digital sum variation (DSV) of the input data must be bounded and kept
as low as possible. Alternatively, a trans-impedance receiver design may
be used, in which case a bounded DSV is not necessary, and the need for
equalisation is also removed.

Appendix A
Fourier series and transforms

Definitions

1. The sine and cosine Fourier series representation of a finite-power, periodic signal $f(t)$ with period $T = 2\pi/\omega$ is:

$$f(t) = A_0 + \sum_{n=1}^{\infty} A_n \cos n\omega t + \sum_{n=1}^{\infty} B_n \sin n\omega t$$

where

$$A_0 = \frac{1}{T} \int_0^T f(t)\, dt = \overline{f(t)}$$

$$\left.\begin{array}{l} A_n = \dfrac{2}{T} \displaystyle\int_0^T f(t) \cos n\omega t\, dt \\[2em] B_n = \dfrac{2}{T} \displaystyle\int_0^T f(t) \sin n\omega t\, dt \end{array}\right\} \quad n = 1, 2, 3, \ldots$$

Note: the integrals may be evaluated over any convenient full period.

2. The exponential (double-sided) representation is:

$$f(t) = \sum_{m=-\infty}^{+\infty} a_m \exp jm\,\omega t$$

where

$$a_m = \frac{1}{T} \int_{-T/2}^{+T/2} f(t) \exp(-jm\,\omega t)\, dt$$

For all integral values of m.
Note that if $f(t)$ is real, as will be the case for a practical signal, $a_m = a_m{}^*$ and the symmetry described in Section 2.3.2 follows. Note also that

$$A_n = 2\,\mathrm{Re}[a_n]$$
$$B_n = -2\,\mathrm{Im}[a_n]$$

3. An alternative trigonometrical (cosine) representation is:

$$f(t) = A_0 + \sum_{n=1}^{\infty} C_n \cos(n\omega t + \phi_n)$$

where

$$C_n = |2a_n| \text{ and } \phi_n = \operatorname{Arg} a_n$$

4. The Fourier Transform of a finite energy signal $g(t)$ is

$$G(f) = \int_{-\infty}^{\infty} g(t) \exp(-j2\pi ft)\, dt$$

This is often denoted $g(t) \leftrightarrow G(f)$.

Example of rectangular pulse height V extending from $-\frac{\tau}{2} < t < \frac{\tau}{2}$

$$G(f) = V \int_{-\tau/2}^{\tau/2} \exp(-j2\pi ft)\, dt$$

$$= V \left[\frac{\exp(-j2\pi ft)}{-j2\pi f} \right]_{-\tau/2}^{\tau/2}$$

$$= \frac{V}{\pi f} \left[\frac{\exp(j\pi f\tau) - \exp(-j\pi f\tau)}{2j} \right]$$

$$= \frac{V \sin \pi f\tau}{\pi f} \quad \text{or} \quad V\tau \left(\frac{\sin \pi f\tau}{\pi f\tau} \right)$$

Some important properties of the Fourier Transform

Suppose that $g(t) \leftrightarrow G(f)$ and $h(t) \leftrightarrow H(f)$. Then,

1 Linearity: $ag(t) + bh(t) \leftrightarrow aG(f) + bH(f)$ where a, b are constants
2 Multiplication: $g(t) \times h(t) \leftrightarrow G(f)*H(f)$ where the symbol * denotes the convolution operation of Appendix B.
3 Convolution: $g(t)*h(t) \leftrightarrow G(f) \times H(f)$.
4 Duality: $G(t) \leftrightarrow g(-f)$. In the important case of an even function for which $g(t) = g(-t)$, then $G(t) \leftrightarrow g(f)$ as illustrated in Figs. 2.24 and 2.25.
5 Modulation: $g(t) \cos(2\pi f_0 t + \phi) \leftrightarrow \frac{1}{2}[G(f-f_0)\exp j\phi + G(f+f_0)\exp(-j\phi)]$

Appendix B
Convolution

Suppose that a continuous system with impulse response $h(t)$ is subject to an input $x(t)$. Consider the input as contiguous, short pulses width δz, and model the pulse at time $z = k\,\delta z$ as an impulse of strength $x(z)\,\delta z$.

Then the output due to this impulse can be written

$$x(z)\,\delta z h(t-z)$$

and the complete output, by superposition, is

$$y(t) = \sum_{k=-\infty}^{\infty} x(z)\,\delta z h(t-z) \quad \text{where } z = k\,\delta z$$

Letting $\delta z \to 0$ we have

$$y(t) = \int_{-\infty}^{\infty} x(z)h(t-z)\,\mathrm{d}z$$

often written simply $y(t) = x(t)*h(t)$.

Note that the convolution integral can also be written

$$h(t)*x(t) = \int_{-\infty}^{\infty} h(z)x(t-z)\,\mathrm{d}z$$

Appendix C

Modelling applications of spreadsheets

Many of the concepts discussed in this book can be illustrated by simple computer simulations. Commercial packages are available for some of them, or simple programs can be written in programming languages such as 'Basic' or 'Pascal'. However, for many topics in digital systems a spreadsheet provides a particularly easy and illuminating demonstration of important techniques.

The following are examples of how spreadsheets can be used to illustrate some of the topics in the text. The details will depend upon the particular spreadsheet package used: it is assumed that readers are already familiar with the package to which they have access. Two different spreadsheet packages are used in the examples below. The zero forcing equaliser is illustrated in 'Excel' on an 'Apple' 'Macintosh' computer, while the other two examples are shown implemented on 'SuperCalc4'.

The reader is encouraged to use these examples as the starting point for experimentation with different parameters and configurations. Other examples of the use of shreadsheets for modelling in engineering will be found in Bissell and Chapman (1989).

A zero forcing equaliser (Section 4.3.4)

Fig. C1 shows a shreadsheet to simulate the zero forcing equaliser of Fig. 4.13. The input (column A) shifts into the first delay stage (column B) then into the second delay stage (column C). The three coefficients (-0.266, 0.866 and 0.204) have been entered in locations E3, E4 and E5 respectively.

The output at each time interval (each line) is calculated in column D from the formula

$$(\text{column A}) \times (\text{E3}) + (\text{column B}) \times (\text{E4}) + (\text{column C}) \times (\text{E5})$$

which appears in the form for line 3 at the top of the display. The effect of changing these coefficients can be easily seen by changing the entries in E3, E4 and E5.

A graphical representation of the equalisation is readily available from the spreadsheet by selecting column A (for the input) and column D (for the output), and is included in Fig. C2.

A self-synchronising scrambler (Section 6.3.1)

Fig. C2 shows the spreadsheet for simulating a three-stage scrambler with tap polynomial $1 + x^{-1} + x^{-3}$, for the input $1010\ldots$ repeating, with the starting state of all 0s. Because this is a binary logic device, all entries are either 0 or 1. Fig. C3 shows the same spreadsheet, displaying the formulae rather than the values in the

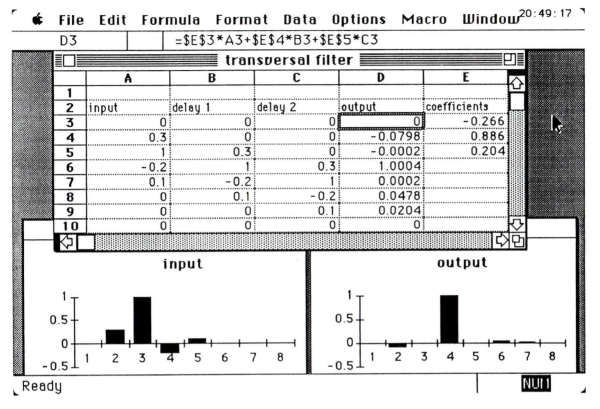

Fig. C1.

	A		B		C		D			E				F		
1	Scrambler 1 + x^-1 + x^-3																
2			scrambler														
3							feed-										
4	in		S1		S2		S3		back				out				
5																	
6	1		0		0		0		0				1				
7	0		1		0		0		1				1				
8	1		1		1		0		1				0				
9	0		0		1		1		1				1				
10	1		1		0		1		0				1				
11	0		1		1		0		1				1				
12	1		1		1		1		0				1				
13	0		1		1		1		0				0				
14	1		0		1		1		1				0				

Fig. C2.

	A	B	C	D	E		F	
1	Scrambler 1 + x^-1 + x^-3							
2			scrambler					
3					feed-			
4	in	S1	S2	S3	back		out	
5								
6	1	0	0	0	B6+D6-2*B6*D6		A6+E6-2*A6*E6	
7	0	F6	B6	C6	B7+D7-2*B7*D7		A7+E7-2*A7*E7	
8	1	F7	B7	C7	B8+D8-2*B8*D8		A8+E8-2*A8*E8	
9	0	F8	B8	C8	B9+D9-2*B9*D9		A9+E9-2*A9*E9	
10	1	F9	B9	C9	B10+D10-2*B10*D10		A10+E10-2*A10*E10	
11	0	F10	B10	C10	B11+D11-2*B11*D11		A11+E11-2*A11*E11	
12	1	F11	B11	C11	B12+D12-2*B12*D12		A12+E12-2*A12*E12	
13	0	F12	B12	C12	B13+D13-2*B13*D13		A13+E13-2*A13*E13	
14	1	F13	B13	C13	B14+D14-2*B14*D14		A14+E14-2*A14*E14	

Fig. C3.

cells. The exclusive-OR logical function is constructed using ordinary arithmetic on the 1s and 0s, using the fact that A exclusive-OR B can be calculated from $A + B - 2AB$.

The output for different starting states can be found by changing the entries in B6, C6 and D6, while the output for a different input sequence will be calculated automatically on changing the entries in column A.

A descrambler can be simulated in a similar way: the two can be combined by putting the descrambler to the right of the scrambler and using the scrambler output as the descrambler input, or by linking two appropriate spreadsheets.

A cyclic redundancy check

The cyclic redundancy check of Figs. 6.12 and 6.13 is simulated in the spreadsheet of Fig. C4. This is similar to the scrambler spreadsheet, but rather more complex.

	A	B	C	D	E	F	G	H	I	J	K	L
1	1 + x + x^3											
2												
3			Transmitter					Receiver				
4												
5		Stage	Stage	Stage	Feed	Out-			Stage	Stage	Stage	Feed
6	Input	one	two	three	back	put	Errors	Input	one	two	three	back
7	1	0	0	0	1	1	0	1	0	0	0	1
8	0	1	1	0	0	0	0	0	1	1	0	0
9	1	0	1	1	0	1	0	1	0	1	1	0
10	0	0	0	1	1	0	0	0	0	0	1	1
11	1	1	1	0	1	1	0	1	1	1	0	1
12	1	1	0	1	0	1	0	1	1	0	1	0
13	0	0	1	0	0	0	0	0	0	1	0	0
14	1	0	0	1	0	1	0	1	0	0	1	0
15	1	0	0	0	1	1	0	1	0	0	0	1
16		1	1	0	0	0	0	0	1	1	0	0
17					0	1	0	1	0	1	1	0
18					0	1	0	1	0	0	1	0
19									0	0	0	0

Fig. C4.

	A	B	C	D	E	F	G	H	I	J	K	L
1	1 + x + x^3											
2												
3		Transmitter						Receiver				
4												
5		Stage	Stage	Stage	Feed	Out-			Stage	Stage	Stage	Feed
6	Input	one	two	three	back	put	Errors	Input	one	two	three	back
7	1	0	0	0	1	1	0	1	0	0	0	1
8	0	1	1	0	0	0	0	0	1	1	0	0
9	1	0	1	1	0	1	0	1	0	1	1	0
10	0	0	0	1	1	0	1	1	0	0	1	0
11	1	1	1	0	1	1	0	1	0	0	0	1
12	1	1	0	1	0	1	0	1	1	1	0	1
13	0	0	1	0	0	0	0	0	1	0	1	1
14	1	0	0	1	0	1	0	1	1	0	0	1
15	1	0	0	0	1	1	0	1	1	0	0	1
16		1	1	0	0	0	0	0	1	0	0	0
17						1	0	1	0	1	0	1
18						1	0	1	1	1	1	0
19									0	1	1	1

Fig. C5.

Both the generation at the transmitter and the check at the receiver are modelled. The following notes explain some of the features of the spreadsheet.

Transmitter

The remainder following polynomial division appears in cells B16, C16 and D16.

The output column F consists of the input data (taken from column A) followed by the calculated remainder.

Receiver

The input to the receiver (column H) is taken from the output of the transmitter, with the option of inserting errors available in column G. Column H is simply the exclusive-OR of columns F and G: a 1 in column G inverts the value from column F.

The remainder following division at the receiver (the error syndrome) is contained in cells I19, J19 and K19.

In Fig. C4, no errors are inserted (column G contains all 0s), so the syndrome is all 0s. Fig. C5 shows the effect of inserting an error (a 1 in cell G10): the syndrome ceases to be all 0s.

Outline answers to numerical exercises

Chapter 2

2.1(a) $2.5\exp(j\omega t + \pi/2) + 2.5\exp(-j\omega t - \pi/2)$; double-sided spectral lines of amplitude 2.5, phase $\pm\pi/2$ at $\pm\omega$.

(b) $0.5[\exp(j\omega t) + \exp(-j\omega t)] + 0.25[\exp(j\omega t + \pi/4) + \exp(-j\omega t - \pi/4)]$; spectral lines of amplitude 0.5, zero phase at $\pm\omega$ and amplitude 0.25, phase $\pm\pi/4$ at $\pm 3\omega$.

2.2(a) $G(0) = 10^{-2}\,\text{Vs}$, zero crossings every $100\,\text{Hz}$; (b) $G(0) = 6.25 \times 10^{-4}\,\text{Vs}$, zero crossings every $8\,\text{kHz}$.

2.3(a) $0.5\,\text{V}$ high extending from -1 to $+1\,\text{ms}$; (b) $10\,\text{V}$ high, extending from -0.025 to $+0.025\,\text{s}$.

Chapter 3

3.1 $V^2/3\,\text{W}$.

3.2(a) zero, as there is no d.c. spectral component; (b) $60\,\text{mV}$.

3.3 Output mean-square voltage $2 \times 10^{-5}\,\text{V}^2$. In other words, the first-order filter lets through approximately half as much noise power again as an ideal filter with same cut-off. Alternatively, an ideal filter with a cutoff of $\pi f_c/2$ would pass the same noise power as the first-order filter.

3.4 The continuous component in each case will be similar to Fig. 3.17(a) but scaled in frequency by a factor of 2 (spectral nulls at integral multiples of $2/T$ hertz). The bipolar *impulse* train has no spectral lines, so neither does the RZ bipolar *pulse* train. The unipolar impulse train has lines at d.c. and all multiples of $1/T$; these appear in the spectrum of the RZ pulse train as appropriately scaled lines at d.c., $1/T$ and frequencies near to side-lobe peaks in the continuous spectral component.

3.5 Noise has zero mean so $v_{\text{rms}} = 67\,\text{mV} = \sigma$. $100\,\text{mV}$ is 1.5σ above mean, so $Q(1.5) \approx 0.065$, say 7%.

3.6 $Q(3.1) = 10^{-3}$, so $10/2\sigma = 3.1$ and $\sigma \approx 1.6\,\text{V}$. Alternatively, reading from Fig. 3.23 $10/\sigma = 16\,\text{dB}$ or a factor of about 6.3, leading to the same result.

Chapter 4

4.2(b) Using coefficients -0.192, 0.958, 0.115 gives equaliser output -0.04, 0.00, 1.00, 0.00, 0.08, 0.01.

4.4(c) Z is located at $(0.5, 1)$ so orthogonal points are found at $(1, -0.5)$ and $(-1, 0.5)$.

4.6 The states all lie on a circle radius $\sqrt{E}$ centred on the origin.

4.7 The states are located at distances 1 and 3 units from the origin at appropriate angles to the horizontal axis.

Chapter 5

5.1 In each case the coding depends upon what has gone before. The possibilities are as follows:

```
Input data:   1 000 1 1 1 0 000 0 000 1 1 0 000 1
AMI either:   + 000 − + − 0 000 0 000 + − 0 000 +
or:           − 000 + − + 0 000 0 000 − + 0 000 −
HDB3:         + 000 − + − 0 00− + 00+ − + − 00− +
or:           − 000 + − + 0 00+ − 00− + − + 00+ −
                                V B  V      B  V
or:           − 000 + − + − 00− + 00+ − + − 00− +
or:           + 000 − + − + 00+ − 00− + − + 00+ −
                          B   V B  V      B  V
```

(V indicates the location of a violation pulse, B a balancing pulse

```
CMI:  − − − + − + − + + + − − + + − + − + − + − + − + − + − + − + − − + + − + − + − + − + − −
or:   + + − + − + − + − − + + − − − + − + − + − + − + − + − + − + + + − − − + − + − + − + + +
```

5.2

Code	Radix	Redundancy	Efficiency
CMI	2	0.5	50%
3B4B	2	0.25	75%
4B3T	3	0.16	84%

Note: 4B3T encodes 4 binary bits to 3 ternary symbols, so the redundancy is calculated:

$$\frac{\text{information per symbol available} - \text{information per symbol used}}{\text{information per symbol available}}$$

$$= \frac{1.58 - 4/3}{1.58} = 0.16$$

5.3

```
Input data: 100     001     100     001     000     001     011     100     000     001
Code:       + − − +  − − + +  + − − +  − − + +  − − + −  − − + +  − + + −  + − − +  + + − +  − − + +
or:         + − − +  − − + +  + − − +  − − + +  + + − +  − − + +  − + + −  + − − +  − − + −  − − + +
```

5.4

| Input data: | 1 | 1 | 0 | 1 | 1 | 0 | 1 | 1 | 0 | 0 | 1 | 1 | 0 | 1 |

CMI: − − + + − + − − + + − + − − + + − + − + − − + + − + − −

or: + + − − − + + + − − − + + + − − − + − + + + − − − + + +

Manchester: − + − + + − − + − + + − − + − + + − + − − + − + + − − +

Manchester
bi-phase
mark: − + − + − − + − + − + + − + − + − − + + − + − + − − + −

or (the
complement). + − + − + + − + − + − − + − + − + + − − + − + − + + − +

5.5

(a) CMI: 3

(b) Manchester code: 2

(c) Manchester bi-phase mark: 2

5.6 The running digital sum of both (a) and (c) exceeds 3, so both must contain errors. (b) does not exceed 3, so may not contain errors.

5.7 CMI has a DSV of 3, Manchester code has 2. Both codes use full-width pulses and for the same information rate have the same signalling rate. For the same level of eye closure therefore, Manchester encoded data can operate on a channel with a high-pass filter characteristic that is 3/2 times higher than that required for CMI encoded data.

Applying the formulae for a 140 Mbits s^{-1} system, the tolerable high-pass cut-off frequencies are 1.5 MHz and 2.2 MHz for CMI and Manchester code respectively.

Chapter 6

6.2

(a) Two-bit repeating pattern: 1010...

(b) Fourteen-bit repeating pattern: 00001101111001...

The scrambler is latched in (a).

6.3

(a) (Bit number 5 is in error): 1000

(b) (Bit number 2 – a parity check – is in error): 1100

(c) (No errors); 1110

6.5 The distances are: A: 5, B: 6, C: 4, D: 5, E: 1, F: 4, G: 6 and H: 5. Thus the surviving paths are E, F, C and either D or H.

Chapter 7

7.1 2.1 dB

Chapter 8

8.1 The tolerance on the 8448 kbits s^{-1} signal is ± 30 ppm, so the range is

$$8\,447\,747 \text{ bits s}^{-1} \text{ to } 8\,448\,253 \text{ bits s}^{-1}$$

Conveyed bit rates at 206 bits per frame is from $(206/848) \times 8\,447\,747$ bits s$^{-1} = 2\,052\,165$ bits s^{-1} to $(206/848) \times 8\,448\,253$ bits s$^{-1} = 2\,052\,288$ bits s^{-1}.

The tolerance on the 2048 Mbits s^{-1} signal is ± 50 ppm, so the range is $2\,047\,898$ bits s^{-1} to $2\,048\,102$ bits s^{-1} and the conveyed data rate is always greater than the incoming data rate.

Chapter 9

9.1 $M \approx 15.2$; $S/N \approx 118$.

References

ADAMS, P.F. and COOK, J.W. (1989). A Review of copper pair local loop transmission systems, *British Telecom Technology Journal*, Vol. 7, no. 2, pp. 17–29, April.

BISSELL, C.C. and CHAPMAN, D.A. (1989). Modelling applications of spreadsheets, *Institution of Electrical Engineers Review*, Jul/Aug., pp. 267–71.

BISSELL, C.C. (1987). A phasor approach to digital filters, *International Journal of Electronic Engineering Education*, Vol. 24, pp. 197–213.

BREWSTER, R.L. (1987). *Telecommunications technology*, Ellis Horwood.

BYLANSKI, P. and INGRAM, D.G.W. (1976). *Digital transmission systems*, Peter Peregrinus.

CCITT (1989). *See* note below.

CLARK, G.C. and CAIN, J.B. (1981). *Error correction coding for digital communications*. Plenum press.

GOLOMB, S.W. (1967). *Shift register sequences*, Holden-Day Inc.

HAMSTRA, J.R. and MOULTON, R.K. (1985). Group coding system for serial data transmission. US Patent No. 4 530 088, July 16, 1985.

JONES, W.B. (1988). *Optical fiber communications systems*, Holt, Rinehart and Winston.

LECHLEIDER, J.W. (1989). Line codes for digital subscriber lines, *IEEE Communications Magazine*, Vol. 27, No. 9, pp. 25–32.

LYNN, P.A. and FUERST, W. (1989). *Introductory digital signal processing with computer applications*, John Wiley.

MACKINNON, D., McCRUM, W. and SHEPPARD, D. (1990). *An introduction to OSI*, Computer Science Press.

MEADE, M.L. and DILLON, C.R. (1991). *Signals and systems*, Chapman and Hall (2nd edition).

OPEN UNIVERSITY (1990). *T322. Digital telecommunications*.

PEEBLES, P.Z. (1987). *Digital communication systems*, Prentice Hall.

SAVAGE, J.E. (1967). Some simple self-synchronising digital data scramblers, *Bell Systems Technical Journal*, Vol. 46, Feb., pp. 449–97.

SCHWARTZ, M. (1990). *Information transmission, modulation and noise*, McGraw Hill International (4th edition).

WATERS, D.B. (1983). Line codes for metallic line systems, *International Journal of Electronics*, Vol. 55, no. 1, pp. 159–69.

CCITT recommendations

New editions of recommendations are published following plenary assemblies of the CCITT every four years. Successive editions are universally referred to by the colour of their covers: the editions current at the time of writing were the 'blue books' published in 1989. Of particular relevance to the material in the book are the following:

Volume III – Fascicle III.4 *General aspects of digital transmission systems; terminal equipments. Recommendations G.700–G.795.*

Volume III– Fascicle III.5 *Digital networks, digital sections and digital line systems. Recommendations G.801–G961.*

Volume III – Fascicle III.7 *Integrated services digital network (ISDN). General structure and service capabilities Recommendations I.110–I.257.*

Volume VIII – Fascicle VIII.1 *Data communications over the telephone network. Series V recommendations.*

Index